AF147164

IFIP Advances in Information and Communication Technology

735

Editor-in-Chief

Kai Rannenberg, Goethe University Frankfurt, Germany

Editorial Board Members

TC 1 – Foundations of Computer Science
Luís Soares Barbosa, University of Minho, Braga, Portugal

TC 2 – Software: Theory and Practice
Jacques Carette, Department of Computer Science, McMaster University, Hamilton, ON, Canada

TC 3 – Education
Arthur Tatnall, Victoria University, Melbourne, Australia

TC 5 – Information Technology Applications
Erich J. Neuhold, University of Vienna, Austria

TC 6 – Communication Systems
Burkhard Stiller, University of Zurich, Zürich, Switzerland

TC 7 – System Modeling and Optimization
Lukasz Stettner, Institute of Mathematics, Polish Academy of Sciences, Warsaw, Poland

TC 8 – Information Systems
Jan Pries-Heje, Roskilde University, Denmark

TC 9 – ICT and Society
David Kreps, National University of Ireland, Galway, Ireland

TC 10 – Computer Systems Technology
Achim Rettberg, Hamm-Lippstadt University of Applied Sciences, Hamm, Germany

TC 11 – Security and Privacy Protection in Information Processing Systems
Steven Furnell, Plymouth University, UK

TC 12 – Artificial Intelligence
Eunika Mercier-Laurent, University of Reims Champagne-Ardenne, Reims, France

TC 13 – Human-Computer Interaction
Marco Winckler, University of Nice Sophia Antipolis, France

TC 14 – Entertainment Computing
Rainer Malaka, University of Bremen, Germany

IFIP Advances in Information and Communication Technology

The IFIP AICT series publishes state-of-the-art results in the sciences and technologies of information and communication. The scope of the series includes: foundations of computer science; software theory and practice; education; computer applications in technology; communication systems; systems modeling and optimization; information systems; ICT and society; computer systems technology; security and protection in information processing systems; artificial intelligence; and human-computer interaction.

Edited volumes and proceedings of refereed international conferences in computer science and interdisciplinary fields are featured. These results often precede journal publication and represent the most current research.

The principal aim of the IFIP AICT series is to encourage education and the dissemination and exchange of information about all aspects of computing.

More information about this series at https://link.springer.com/bookseries/6102

Denis Cavallucci · Stelian Brad ·
Pavel Livotov
Editors

World Conference of AI-Powered Innovation and Inventive Design

24th IFIP WG 5.4 International TRIZ Future Conference, TFC 2024
Cluj-Napoca, Romania, November 6–8, 2024
Proceedings, Part I

 Springer

Editors
Denis Cavallucci (ID)
INSA Strasbourg
Strasbourg, France

Stelian Brad (ID)
Technical University of Cluj-Napoca
Cluj-Napoca, Romania

Pavel Livotov (ID)
Offenburg University
Offenburg, Germany

ISSN 1868-4238 ISSN 1868-422X (electronic)
IFIP Advances in Information and Communication Technology
ISBN 978-3-031-75918-5 ISBN 978-3-031-75919-2 (eBook)
https://doi.org/10.1007/978-3-031-75919-2

© IFIP International Federation for Information Processing 2025

This work is subject to copyright. All rights are solely and exclusively licensed by the Publisher, whether the whole or part of the material is concerned, specifically the rights of translation, reprinting, reuse of illustrations, recitation, broadcasting, reproduction on microfilms or in any other physical way, and transmission or information storage and retrieval, electronic adaptation, computer software, or by similar or dissimilar methodology now known or hereafter developed.
The use of general descriptive names, registered names, trademarks, service marks, etc. in this publication does not imply, even in the absence of a specific statement, that such names are exempt from the relevant protective laws and regulations and therefore free for general use.
The publisher, the authors and the editors are safe to assume that the advice and information in this book are believed to be true and accurate at the date of publication. Neither the publisher nor the authors or the editors give a warranty, expressed or implied, with respect to the material contained herein or for any errors or omissions that may have been made. The publisher remains neutral with regard to jurisdictional claims in published maps and institutional affiliations.

This Springer imprint is published by the registered company Springer Nature Switzerland AG
The registered company address is: Gewerbestrasse 11, 6330 Cham, Switzerland

If disposing of this product, please recycle the paper.

Preface – TRIZ Future Conference 2024

The 2024 edition of the TRIZ Future Conference (TFC 2024) marked a significant milestone not only in the history of the conference but also in the field of inventive problem-solving. This year, the conference introduced the new acronym TRAI, which stands for 'TRIZ and AI'. This marks the beginning of a new era in the conference's history, reflecting a pivotal shift towards integrating digital creativity and AI-driven solutions in TRIZ methodologies.

Over the course of its 24 editions, the TRIZ Future Conference has continually evolved, and the introduction of TRAI represents the next stage in this evolution. TRAI acknowledges the growing importance of digital tools and AI in fostering creativity and solving complex problems. This transformation is crucial as industries worldwide increasingly rely on these technologies to drive innovation.

This year's conference received 72 submissions, all of which underwent a rigorous peer review process. Each paper was reviewed by two or three members of the scientific committee through a double-blind review process. The review ensured that only high-quality and relevant contributions were selected for presentation and publication in this volume. In total, 42 full papers were accepted.

Special attention was given to contributions that embraced the new TRAI framework, exploring the synergies between TRIZ and AI. Several papers co-authored by committee members were reviewed independently to maintain the integrity of the process.

The papers in this volume are organized into five thematic sections: AI-Driven TRIZ and Innovation; Sustainable and Industrial Design with TRIZ; Digital Transformation, Industry 4.0, and Predictive Analytics; Interdisciplinary and Cognitive Approaches in TRIZ; and Customer Experience and Service Innovation with TRIZ. These sections demonstrate the wide-ranging applications of TRIZ and the new opportunities presented by the TRAI initiative.

We extend our deepest gratitude to all authors, reviewers, and participants who contributed to the success of TFC 2024. Our special thanks go to the supporting institutions for their invaluable assistance in making this event possible.

September 2024

Denis Cavallucci
Stelian Brad
Pavel Livotov

Organization

General Chair

Stelian Brad   Technical University of Cluj-Napoca, Romania

Scientific Chair

Denis Cavallucci   INSA Strasbourg, France

Program Committee

Denis Cavallucci	INSA Strasbourg, France
Pavel Livotov	Offenburg University of Applied Sciences, Germany
Stelian Brad	Technical University of Cluj-Napoca, Romania

Local Organizing Committee

Paulina Mitrea	Technical University of Cluj-Napoca, Romania
Emilia Brad	Technical University of Cluj-Napoca, Romania
Calin Neamtu	Technical University of Cluj-Napoca, Romania
Ionut Chis	Technical University of Cluj-Napoca, Romania
Anca Stan	Technical University of Cluj-Napoca, Romania
Vlad Trifan	Technical University of Cluj-Napoca, Romania
Bogdan Balog	Technical University of Cluj-Napoca, Romania
Alexandru Cirlejan	Technical University of Cluj-Napoca, Romania
Cosmin Muresan	Technical University of Cluj-Napoca, Romania
Dragos Bartos	Technical University of Cluj-Napoca, Romania
Eyas Deeb	Technical University of Cluj-Napoca, Romania
Andrei Kelemen	Cluj IT Cluster, Romania

Scientific and Professional Committee

Robert Adunka	TRIZ Consulting Group GmbH, Germany
Nuri Başoğlu	Izmir Institute of Technology, Türkiye
Robert Bembenik	Warsaw University of Technology, Poland
Rachid Benmoussa	Cadi Ayyad University, Marrakesh, Morocco
Tiziana Bertoncelli	ANSYS Germany, Germany
Stelian Brad	Technical University of Cluj-Napoca, Romania
Denis Cavallucci	INSA Strasbourg, France
Leonid Chechurin	Lappeenranta-Lahti University of Technology, Finland
Hicham Chibane	INSA Strasbourg, France
Jerzy Chrząszcz	Warsaw University of Technology, Poland

Michał Chwesiuk	Warsaw University of Technology, Poland
Marin Iuga	Intertechnica, Romania
Delia Mitrea	Technical University of Cluj-Napoca, Romania
Emilia Brad	Technical University of Cluj-Napoca, Romania
John Cooke	CoCatalist Ltd./Water-Trak Ltd., UK
Amadou Coulibaly	INSA Strasbourg, France
Alexander Czinki	TH Aschaffenburg University of Applied Sciences, Germany
Marco A. De Carvalho	Universidade Tecnológica Federal do Paraná, Brazil
Roland De Guio	INSA Strasbourg, France
Christoph Dobrusskin	Philips, The Netherlands
Sébastien Dubois	INSA Strasbourg, France
Kalle Elfvengren	Lappeenranta-Lahti University of Technology, Finland
Oleg Feygenson	MATRIZ Official, Samsung Electronics, South Korea
Kai Hiltmann	Coburg University of Applied Sciences, Germany
Rémy Houssin	University of Strasbourg, France
Mariusz Kaleta	Warsaw University of Technology, Poland
Karl Koltze	Hochschule Niederrhein, Germany
Narayanan Kulathuramaiyer	Universiti Malaysia Sarawak (UNIMAS), Malaysia
Krzysztof Kulpa	Warsaw University of Technology, Poland
Jane Labadin	Universiti Malaysia Sarawak (UNIMAS), Malaysia
Jan-Charles Lamirel	LORIA, University of Lorraine, France
Claus Lang-Koetz	Pforzheim University, Germany
Ralf Laue	University of Applied Sciences Zwickau, Germany
Pavel Livotov	Offenburg University of Applied Sciences, Germany
Lorenzo Maccioni	Free University of Bozen-Bolzano, Italy
Nicolas Maranzana	École Nationale Supérieure d'Arts et Métiers Paris, France
Oliver Mayer	Bayern Innovativ GmbH, Germany
Horst Nähler	c4pi/TRIZ-Akademie, Germany
Toru Nakagawa	Osaka Gakuin University, Japan
Robert Nowak	Warsaw University of Technology, Poland
Piotr Pałka	Warsaw University of Technology, Poland
Alain Riwan	Centre d'intégration Nano-INNOV, France
Federico Rotini	University of Florence, Italy
Davide Russo	University of Bergamo, Italy
Justus Schollmeyer	University of Bremen, Germany
Vladimir Sizikov	SIEGENIA Group, Germany
Christian Spreafico	University of Bergamo, Italy
Christian M. Thurnes	Hochschule Kaiserslautern, Germany
Jarosław Turkiewicz	Warsaw University of Technology, Poland
Jörg Vetter	J. Vetter-S3-consulting, Germany
Julian Vincent	Heriot-Watt University, UK

Contents – Part I

Contents – Part II

Digital Transformation, Industry 4.0, and Predictive Analytics

Interdisciplinary and Cognitive Approaches in TRIZ

Customer Experience and Service Innovation with TRIZ

AI-Driven TRIZ and Innovation

LLM-Based Extraction of Contradictions from Patents

Stefan Trapp[✉] and Joachim Warschat

University of Hagen, Universitätsstraße 47, 58097 Hagen, Germany
stefan.trapp@fernuni-hagen.de

Abstract. Already since the 1950s TRIZ shows that patents and the technical contradictions they solve are an important source of inspiration for the development of innovative products. However, TRIZ is a heuristic based on a historic patent analysis and does not make use of the ever-increasing number of latest technological solutions in current patents. Because of the huge number of patents, their length, and, last but not least, their complexity there is a need for modern patent retrieval and patent analysis to go beyond keyword-oriented methods. Recent advances in patent retrieval and analysis mainly focus on dense vectors based on neural AI Transformer language models like Google BERT. They are, for example, used for dense retrieval, question answering or summarization and key concept extraction. A research focus within the methods for patent summarization and key concept extraction are generic inventive concepts respectively TRIZ concepts like problems, solutions, advantage of invention, parameters, and contradictions. Succeeding rule-based approaches, finetuned BERT-like language models for sentence-wise classification represent the state-of-the-art of inventive concept extraction. While they work comparatively well for basic concepts like problems or solutions, contradictions − as a more complex abstraction − remain a challenge for these models. Even PaTRIZ, the latest and complicated multistage approach to extract contradictions, delivers only mixed results. This paper goes one step further, as it presents a method to extract TRIZ contradictions from patent texts based on Prompt Engineering using a generative Large Language Model (LLM), namely OpenAI's GPT-4. The existing annotated patent dataset "PaGAN" is used to demonstrate the LLM-capabilities for extracting TRIZ contradictions from the section "State-of-the-Art" of USPTO patents. Contradiction detection, sentence extraction, contradiction summarization, parameter extraction and assignment to the 39 abstract TRIZ engineering parameters are all performed in a single prompt using the LangChain framework. Our results show that "off-the-shelf" GPT-4 is a serious alternative to PaTRIZ. Comparing the text similarity of the GPT-4 extractions with the annotated sentences from PaGAN we reach a high F1-value of 0.93 using the BERTScore metric.

Keywords: AI · BERT · Classification · Contradiction · Finetuning · GPT-4 · Information Extraction · Information Retrieval · Inventive Concept · LangChain · Large Language Model (LLM) · NLP · OpenAI · Prompt Engineering · Summarization · Transformer · TRIZ

© IFIP International Federation for Information Processing 2025
Published by Springer Nature Switzerland AG 2025
D. Cavallucci et al. (Eds.): TFC 2024, IFIP AICT 735, pp. 3–19, 2025.
https://doi.org/10.1007/978-3-031-75919-2_1

1   Introduction

Innovations are a key success factor for companies to compete in the age of globalization. Emerging technologies are an important driver for innovative products as they provide new or improved solutions for customer needs. But as the technical knowledge published worldwide grows rapidly and technologies become increasingly complex, it gets harder for companies to identify attractive technological opportunities especially outside their own domain. This implies the risk of missing important technological developments and of falling behind in competition [1].

Because 85 to 90 percent of the worldwide technical knowledge can be found in patents, they are an extensive and cross-domain knowledge resource for inventive technological solutions [2]. This potential of patents for improved innovation results was already discovered by Altshuller in the 1950s, when he developed TRIZ, the theory of inventive problem solving [3]. Back then patent search engines did not yet exist, but Altshuller's manual patent analysis revealed two important insights. First, the resolution of contradictions between system parameters often creates highly innovative solutions. Second, many high-grade innovations are based on the cross-domain transfer of already existing solutions. Until today TRIZ remains a classic approach promising technical solutions of higher innovativeness than general-purpose creativity techniques. TRIZ is based on a historic patent analysis, but it does not make use of the ever-increasing number of latest technological solutions in current patents [4].

Nowadays, patent search engines support the retrieval of patents (possibly) containing technological solutions of interest. However, especially when it comes to cross-domain search, the widespread keyword-based lexical patent search suffers from the vocabulary mismatch problem: due to different terms, synonyms or a completely different jargon between domains, the word-overlap may only be very little, resulting in unsatisfactory search results.

A possible solution to simultaneously make use of the TRIZ findings and to overcome the problems of conventional lexical patent search is the extraction of inventive (key) concepts from patents and their subsequent use for patent retrieval. Generic inventive concepts present in all patents are "problems", "advantages of the invention" and "solutions" [5]. TRIZ adds the entities "parameter" and "contradiction" [4]. Among these inventive concepts contradictions are the most complex concept, and, according to TRIZ, also the most important one to find highly innovative solutions, if a solution is presented to overcome a contradiction. If the contradictions addressed by patents are extracted automatically, they can, together with references to the source patents, be stored in a dedicated retrieval index. This way, engineers can perform a cross-domain search for patents addressing contradictions similar to a contradiction they are currently facing in the innovation process. The solutions presented in the retrieved patents may be transferable to the engineer's own domain resulting in innovative problem solutions.

Existing solutions for inventive concept extraction rely either on rule-based natural language processing (NLP) or, lately, on Artificial Intelligence (AI) in the form of pre-trained neural language models like Bidirectional Encoder Representations from Transformers (BERT) [6] developed by Google. After additional finetuning with annotated data, BERT and its derivatives are typically used for sentence-wise classification to extract inventive concepts. However, the automatic detection of contradictions in patent text remains a challenge.

Generative Large Language Models (LLMs), like OpenAI's omnipresent GPT (Generative Pre-trained Transformer) [7], are the latest trend in pre-trained neural language models. With their improved capabilities to understand and generate text LLMs often deliver satisfying results without effortful finetuning and simply via instructions and input in plain natural language (prompts and Prompt Engineering). Thus, they might also be an alternative to smaller models like BERT for inventive concept extraction.

The aim of this paper is to analyze the extraction of technical contradictions from patents in English language using "out-of-the-box" LLMs and Prompt Engineering. The LLM results are compared with the state-of-the-art, namely sentence classification with finetuned BERT derivatives and complicated multi-stage approaches based on the same.

2 Background and Related Work

2.1 Patents and Their Inventive Concepts

Over 20 million patents exist worldwide. Their average text length is around 10,000 words. In addition to their length, their complicated writing style, with the intention to hide the intellectual property as much as possible, causes the vocabulary mismatch problem for keyword-oriented patent search. Contextual text embeddings based on neural networks avoid the dependency on words and word-overlap as they code the text's meaning. Therefore, they are a potential workaround for the vocabulary mismatch problem, which hinders particularly cross-domain search. The latest generation of such neural networks are transformer architectures like Google BERT [8].

According to the guidelines of the World Intellectual Property Organization (WIPO), patents should cover the problem to be solved by the invention, the solution to overcome the problem and the advantageous effects of the invention [5]. Thus, these three generic inventive concepts can be expected to be present in all patents. The problems or technical contradictions addressed in a patent are often presented in its "Background" section describing the state-of-the-art [9].

2.2 TRIZ and Its Inventive Concepts

The probably most famous and widespread tool of the TRIZ heuristic is the contradiction matrix for solving technical conflicts between parameters. The TRIZ-user must abstract the targeted and the deteriorated parameters of the problematic contradiction to one of the 39 TRIZ engineering parameters to select a cell of the contradiction matrix, thus obtaining a subset of the 40 generic TRIZ innovative principles as guideline for further solution search. These innovative principles proved to be successful for this class of problems in Altshuller's historic patent analysis and can be used to look for solutions adaptable to one's own problem [10].

Parameters and contradictions can be seen as further generic inventive concepts in addition to problems, solutions, and advantageous effects. Contradictions between system parameters are a specific type of problems. According to TRIZ, overcoming contradictions promises excellent chances for highly innovative solutions.

Later developments based on TRIZ principles, like the Inventive Design Method (IDM) formalize TRIZ concepts through ontologies, enabling automated text analysis. IDM defines TRIZ contradictions as a set of two evaluation parameters (corresponding to the engineering parameters in the rows and columns of the TRIZ contradiction matrix) and possibly a third parameter, the action parameter, corresponding to the "physical contradictions" of classic TRIZ [4].

2.3 Extraction of Inventive Concepts from Patents

Many approaches targeting the extraction of inventive concepts from patents rely on rule-based NLP (syntactics of sentences and signaling markers) to retrieve entities and to link them to each other [11, 12]. However, a problem with rule-based methods is, that it is difficult to create a precise and exhaustive set of rules, especially given the complex language used in patents. For instance, the support of n-grams is crucial: "Device for stopping two-wheeled vehicles" is an example of a subject that consists of five words. Simply extracting the word "Device" does not convey the original meaning. A second problem is, that the code used for extraction is often not being disclosed. Furthermore, unspecific rules may apply to almost every sentence, with the consequence that the desired text summary focusing on inventive concepts is only partially achieved.

Heller and Warschat show that BERT outperforms static word embeddings such as Word2vec and GloVe in the sentence-wise classification of problems and solutions [13]. Similar to Heller and Warschat, Giordano et al. fine-tune a BERT-for-Patents model [14] for classifying patent sentences into the four classes "problem", "solution", "advantage (of the invention)" or "other sentences" [5]. However, while sentence-wise classification works well for simple inventive concepts like problems, the extraction of contradictions, as a more complex concept, remains a challenge as shown below.

Guarino et al. use two BERT-based binary sentence classifiers in their "SummaTRIZ" approach to extract sentences from patents that define contradictions [9]. "SummaTRIZ" was later supplemented with further NLP steps to form the "PaTRIZ" approach [15–17]. "PaTRIZ" is probably the most advanced contradiction extraction method identified in the state-of-the-art [15–17]. It is a complicated four-stage approach. Stage (1) is a "yes/no" classifier for the presence of contradictions in the respective patent. Stage (2) uses (like "SummaTRIZ") two binary BERT classifiers to extract those sentences that describe a contradiction. The first sentence classifier identifies sentences with the parameter to be improved ("first part of the contradiction"), the second identifies those with the deteriorating parameter ("second part of the contradiction"). The separate classification aims to find contradictions in which both components are contained in the same sentence [17]. Stage (3) is a Conditional Random Field (CRF) for extracting the parameters from the "contradiction sentences". In stage (4) an MPNet transformer classifies the extracted (concrete) parameters as one of the 39 abstract engineering parameters of the (classic) TRIZ contradiction matrix. This classification is based on the semantic text similarity with the textual description of the TRIZ parameters.

Despite its complexity and high finetuning effort the best PaTRIZ overall model only achieves a precision of 0.33, i.e. out of three contradictions extracted, only one matches to an expert-annotated contradiction. The classifier stage (2) alone reaches F1-values of 0.37 for the "first part" resp. F1 = 0.65 for the "second part" [17].

2.4 Finetuning Versus Prompt Engineering and In-Context-Learning

Finetuning is an approach for improving the results of pre-trained language models on a downstream task using an additional task-specific dataset. In contrast to other techniques like tuning the prompts for querying the model (Prompt Engineering), finetuning adapts the model weights (parameters) of the pre-trained language model. Supervised learning of neural models requires comprehensive training data, but this data is often difficult to obtain or difficult to annotate. Specifically in the patent domain there is a lack of such data [18, 19].

Larger language models usually achieve significant performance improvements compared to smaller models, which can be described by so-called scaling laws [20, 21]. BERT with its 110 million (BERT-base) or 340 million (BERT-large) parameters is nowadays only considered a "medium-sized language model" [22] and is often used for natural language understanding tasks like text classification and keyword extraction. Since BERT, as a masked language model, was not designed to be generative (encoder architecture), it is considered less suitable for language generation tasks compared to generative models such as GPT. However, with its 175 billion parameters, GPT-3 is almost 1,600 times larger than BERT-base. With the successor GPT-4, the size has increased sixfold again, to around one trillion parameters [23]. Currently only GPT and comparable models with their greatly improved abilities in terms of language understanding, language generation, generalization, problem solving and the derivation of logical conclusions (logical reasoning) are considered as "Large Language Models" (LLMs). [24, 25].

For medium-sized language models such as BERT and its derivatives, finetuning is usually essential to achieve satisfactory results. On the other side, LLMs can be finetuned, too (with appropriately powerful hardware), but in many cases they deliver satisfying results without effortful finetuning and simply via instructions and input in plain natural language (prompts and Prompt Engineering). Prompts and Prompt Templates allow Few-shot learning, i.e. learning based on a selection of solved examples, without adjusting model parameters. This approach is also known as in-context learning (ICL). In many cases the thorough design of so-called zero-shot prompts can achieve comparable performance even without inserted examples [26, 27].

With their avoidance of finetuning and the intuitive codeless interaction via prompts LLMs might be an interesting alternative to smaller language models like BERT for inventive concept extraction and specifically for technical contradictions.

3   Proposed Method

Our idea is to improve the use of the knowledge contained in millions of patents for innovative problem solving through new search capabilities. To achieve this, contradictions addressed in patents should be extracted independently of specific search queries and stored in a search index along with their contextual text embeddings (ingestion in a vector database). Later, when a solution is sought for a contradiction that has arisen in the innovation process, the indexed contradictions can be specifically used for a similarity search (Fig. 1). In this publication, we focus solely on the extraction of contradictions, shown with gray filling in Fig. 1. This chapter describes a new, LLM-based method to extract technical contradictions from patent texts.

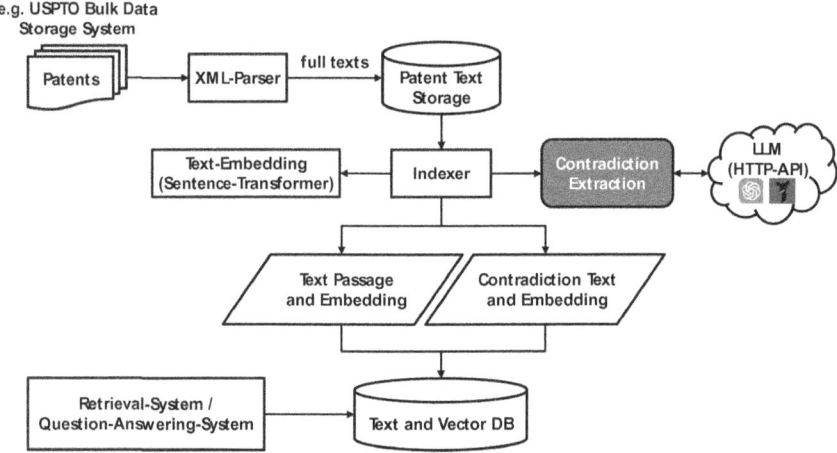

Fig. 1. Big Picture: Improved Patent Search with Dense Retrieval of Contradictions

3.1 Dataset and Data Preparation

In this paper we reuse a patent corpus (called "PaGAN dataset" hereafter) provided by a research team of the Institut national des sciences appliquées de Strasbourg (INSA). It contains 3,200 English-language patents from the database of the United States Patent and Trademark Office (USPTO). Half of these patents (1,600) contain expert-annotated TRIZ contradictions, usually consisting of at least two sentences ("first part of the contradiction" and "second part of the contradiction"). These sentences were extracted from the state-of-the-art section of the patents. The other half of the corpus is supposed to contain no contradictions [28]. Because this dataset with expert-annotated contradictions is the only one we know of; the quantitative evaluation of our approach is currently limited to the possibilities that this dataset offers.

The PaGAN dataset is delivered in multiple text files per patent. Each patent section is stored in a separate file, plus (if applicable) a text file for the expert-annotated sentences describing the contradiction. This format makes the data access cumbersome and slow for batch-use in combination with language models. Thus, a data preparation step was performed. The PaGAN text files were "pickled" into a single Python object serialization file using the Pandas library (https://pandas.pydata.org/).

3.2 LLM-Based Contradiction Extraction with LangChain and GPT-4

LangChain is a Python framework to create LLM-powered applications [29]. It can be used in cloud-based development environments like Google Colab [30]. Colab allows the easy execution of Python code as Jupyter Notebooks on GPUs with high computing capacities, thus the environment is particularly suitable for training and executing neural language models.

LangChain is prepared to use various open-source or commercial LLMs via their Application Programming Interface (API). One of these LLMs is OpenAI's GPT. In its version GPT-4 it is OpenAI's most advanced system and benchmarks consider it as the probably most powerful LLM currently available [7, 31, 32]. We use GPT-4 Turbo release *gpt-4-1106-preview*, presented by OpenAI on 6 November 2023, to process the following prompt for contradiction extraction. LangChain's API-agnostic role "System" sets the objectives the AI should follow. We use it to explain the term "TRIZ contradiction" and give definitions for the 39 engineering parameters of the classic TRIZ contradiction matrix. The LangChain role "Human" sends messages from the perspective of the human user. We here define the tasks to be performed by the AI, request an answer in JSON (JavaScript Object Notation), specify its exact format, and, finally, present the text to be analyzed. The JSON format makes it easy to parse the AI response during the intended ingestion into a vector database for later patent search. For the sake of brevity, the TRIZ engineering parameters 2 to 38 are omitted in the prompt below.

System:
As a worldclass patent and technology expert you are extracting TRIZ contradictions from text passages of patents.
A TRIZ contradiction is indicated by at least 2 conflicting parameters, attributes or properties that cannot be optimized simultaneously.
These parameters are called 'evaluation parameters'. If one of these parameters is improved, the other one deteriorates and vice versa.
Possibly a third parameter is part of the contradiction, which influences both 'evaluation parameters' in opposite directions. This optional third parameter is called 'action parameter'. The 'action parameter' must be maximized to improve one of the 'evaluation parameters' while it must be minimized to improve the other.
Each 'evaluation parameter' can be classified as one of the 39 abstract TRIZ-parameters of the TRIZ contradiction matrix given below.
'1 -   Weight of Moving Object': The mass of the object, in a
        gravitational field. The force that the body exerts on
        its support or suspension.
// ...TRIZ-parameters 2-38

'39 - Productivity': The number of functions or operations
        performed by a system per unit time. The time for a unit
        function or operation. The output per unit time, or the
        cost per unit output.

Human:
Perform the following tasks on the given text:
1. Decide if the text mentions at least one TRIZ contradiction.
    Your answer must be either 'True' or 'False'. Answer only
    'True', if the conflicting parameters are apparent. 'False'
    otherwise.

If the answer to task 1 is 'True', perform additionally the following tasks per contradiction mentioned:
2. Quote without any modification the sentences containing the parameters which need to be improved respectively involved in the initial problem. Quote a maximum of 6 sentences. Order them by decreasing importance. Avoid duplicate sentences for this task and task 3 below.
3. Quote without any modification the sentences containing the draw backs of the prior art solutions. Quote a maximum of 6 sentences. Order them by decreasing importance. Avoid duplicate sentences for this task and task 2 above.
4. Summarize the TRIZ contradiction in one sentence WITHOUT any introductory phrase like 'The TRIZ contradiction is that ...'. Include values and units for the parameters, if given.
5. Analyze the parameters of the TRIZ contradiction. Return the 'evaluation parameters' and, if mentioned, the 'action parameter'.
6. Assign each of the 'evaluation parameters' to one of the 39 TRIZ-parameters of the TRIZ contradiction matrix. Quote solely the text in inverted commas from their description, nothing else.

Format your response in a structured JSON-like format, as follows:

```
{"Contradictions": [{
  "is_contradiction": "True/False",
  "target_sentences": "['Only if 'is_contradiction' is True:
  quote without any modification the sentences containing the
  parameters which need to be improved respectively involved in
  the initial problem. Quote a maximum of 6 sentences.']",
  "drawback_sentences": "['Only if 'is_contradiction' is True:
  quote without any modification the sentences containing the
  drawbacks of the prior art solutions. Quote a maximum of 6
  sentences.']",
  "summary": "summary, only if 'is_contradiction' is True",
  "evaluation_parameter_1": "parameter, only if
  'is_contradiction' is True",
  "evaluation_parameter_2": "parameter, only if
  'is_contradiction' is True",
  "triz_parameter_1": "TRIZ parameter, only if
  'is_contradiction' is True",
  "triz_parameter_2": "TRIZ parameter, only if
  'is_contradiction' is True",
  "action_parameter": "parameter, only if 'is_contradiction' is
  True and the action_parameter exists",
},
// ... additional contradictions, if present
]}
```

Return the JSON code only.

The text to analyze is: ...

4   Evaluation

The results of the LLM-based contradiction extraction are evaluated in this chapter in a qualitative and a quantitative way. For the qualitative evaluation we present two patents as case study. The quantitative evaluation relies on the portion of correctly extracted sentences compared with the PaGAN annotations and on the semantic text similarity measured with the F1-value of the BERTScore metric.

4.1   Case Study

The Patents US06938300 "Wheel assembly for a stroller" and US07543344 "Cover for a heating blanket" are used as cases to evaluate the proposed LLM contradiction extraction. The section's "State-of-the-Art" content from US06938300 is shown below. Contradiction sentences according to PaGAN dataset are annotated in **bold**:

---

1. This invention relates to a stroller, and more particularly to a wheel assembly for a stroller, which includes a single wheel. 2. Referring to 1 and 2, a conventional stroller 1 is shown to include a stroller frame 10 with four legs 101, two front wheel assemblies 11 mounted respectively to two of the legs 101, and two rear wheel assemblies 12 mounted respectively to the other two of the legs 101. 2 is a bottom view of one of the front wheel assemblies 11. Each of the front wheel assemblies 11 includes two front wheels 13, a wheel seat 14 disposed between the front wheels 12, and a wheel axle 15 extending through the front wheels 13 and the wheel seat 14. Each of the wheel seats 14 has a leg-connecting portion 141 sleeved rotatably on a respective one of the legs 101, and an axle-connecting portion 142, within which the wheel axle 15 is journalled. To enhance the comfort of the baby carried on the stroller 1, a vibration-absorbing device not shown can be disposed within the axle-connecting portion 142 of each of the wheel seats 14. When it is desired to push the stroller 1 to advance along a straight path, a forward force A is applied to the stroller frame 10. In each of the front wheel assemblies 11, since the forward force A is located midway between two frictional forces B that are generated between the ground and the front wheels 13 and since the direction of the forward force A is parallel to those of the frictional forces B, the stroller 1 can advance along a straight path 16. **When the stroller 1 moves over a lawn or uneven road surfaces, it is necessary for the stroller wheels to have a large diameter so as to ensure the comfort of the baby. However, if each of the front wheel assemblies 11 has two large-diameter front wheels 13, the total volume and weight of the stroller 1 will increase significantly so that it is difficult to push the stroller 1.**

Feeding this patent section into the LLM *gpt-4–1106-preview* using LangChain and the prompt described in Sect. 3 leads to the following AI-answer in JSON:

```
{
  "Contradictions": [
    {
      "is_contradiction": "True",
      "target_sentences": [
        "When the stroller 1 moves over a lawn or uneven road
        surfaces, it is necessary for the stroller wheels to
        have a large diameter so as to ensure the comfort of the
        baby."
      ],
      "drawback_sentences": [
        "However, if each of the front wheel assemblies 11 has
        two large-diameter front wheels 13, the total volume and
        weight of the stroller 1 will increase significantly so
        that it is difficult to push the stroller 1."
      ],
      "summary": "Large-diameter wheels are needed for baby com-
      fort on uneven surfaces, but increase the stroller's vol-
      ume and weight, making it difficult to push.",
      "evaluation_parameter_1": "comfort of the baby",
      "evaluation_parameter_2": "total volume and weight of the
stroller",
      "triz_parameter_1": "'Convenience of Use'",
      "triz_parameter_2": "'Weight of Moving Object'",
      "action_parameter": "diameter of the stroller wheels"
    }
  ]
}
```

In this case both sentences of the contradiction are extracted correctly. The GPT-4-generated summary covers all aspects of the contradiction. Evaluation parameters, action parameter and the assigned generic TRIZ engineering parameters are also correct. Altogether GPT-4's answer can be deemed as very good.

As a second case we use the patent US07543344 ("Cover for a heating blanket"). The content of the "State-of-the-Art" section is omitted here for brevity. It consists of a total of 3008 characters in 21 partially very long sentences. GPT-4's result is shown below.

```
{
  "Contradictions": [
    {
      "is_contradiction": "True",
      "target_sentences": [
        "FAW has become the standard of care for preventing and
        treating the hypothermia caused by anesthesia and sur-
        gery.",
        "There is a need for heating blanket covers that are
        easily manipulated for placing a heating blanket therein
        and do not compromise the comfort of a person over which
        the blanket is placed."
      ],
      "drawback_sentences": [
        "Although FAW is clinically effective, it suffers from
        several problems including: a relatively high price; air
        blowing in the operating room, which can be noisy and
        can potentially contaminate the surgical field; and the
        inflatable blanket is relatively bulky over the patient,
        at times even obscuring the view of the surgeon.",
        "Moreover, the low specific heat of air and the rapid
        loss of heat from air requires that the temperature of
        the air, as it leaves the hose, be dangerously highin
        some products as high as 45 C.",
        "This creates significant dangers for the patient.",
        "Cleaning is not only time consuming during the rapid
        turnover of the operating room after each case, but the
        labor for the cleaning is also expensive."
      ],
      "summary": "The need to prevent hypothermia in surgical
      patients with FAW conflicts with the drawbacks of high
      price, noise, potential contamination, bulkiness, and
      safety hazards.",
      "evaluation_parameter_1": "comfort of a person over which
      the blanket is placed",
      "evaluation_parameter_2": "safety hazards",
      "triz_parameter_1": "'33 - Convenience of Use'",
      "triz_parameter_2": "'30 - Harmful Factors Acting on Ob-
ject'",
      "action_parameter": "temperature of the air"
    }
  ]
}
```

According to the PaGAN dataset 6 sentences are part of the contradiction. GPT-4 also identifies 6 sentences, but only 2 of them match with PaGAN. This result does not look very promising at a first glance. However, content-related (human) analysis shows that GPT-4's results are indeed plausible. The pros and cons of "forced-air warming" (FAW) are described in all 6 sentences. Furthermore, summary and all parameters including the assigned TRIZ engineering parameters seem correct.

The case study shows that the extraction (classification) of sentences for contradiction-identification may not be possible as precisely as desired, because of the complexity and fuzziness of patent language. GPT-4 as an example of state-of-the-art

LLMs shows promising abilities not only to extract such sentences, but also to summarize technical contradictions, to identify their parameters and to assign them to the generic TRIZ engineering parameters.

4.2 Quantitative Evaluation

Out of the 1,600 patents annotated in the PaGAN dataset as patents **with** contradictions, an arbitrary selection of 100 patents was fed into the LLM *gpt-4-1106-preview* using *LangChain*. Table 1 shows the results. In each of the 100 patents GPT-4 found at least one contradiction. In one of the patents two contradictions were identified.

The average number of annotated sentences in the PaGAN dataset is 3.81. GPT-4 identifies 2.69 or 71% of these correctly (*Recall* = 0.71), 1.12 (29%) are not found (false negative). In 73 of the 100 patents GPT-4 found either all annotated sentences (36) or missed at most one (37). In 17 patents two sentences are missed and in the remaining 10 patents GPT-4 misses more than two sentences. For only two of the 100 patents the number of correctly extracted sentences is zero. On average 3.19 sentences are false positive, i.e. identified by GPT-4, but not annotated in PaGAN. Consequently, *Precision* is 0.46 and F1 is 0.56. The best PaTRIZ part models for sentence classification reach $F1_1 = 0.37$ ("first part" based on 2,276 positive sentences) resp. $F1_2 = 0.65$ ("second part" based on 3,709 negative sentences) [17]. The weighted average of $F1_1$ and $F1_2$ is 0.54. Based on our sample of 100 patents it can thus be concluded that a zero-shot *gpt-4–1106-preview* (F1 = 0.56) is at least on par with (if not slightly better than) the best performing fine-tuned sentence classifiers from PaTRIZ (F1 = 0.54).

As the PaGAN dataset provides the contradiction-involved sentences only "as is" and not a summary of the contradiction, these sentences are the basis for our quantitative similarity evaluation. We use the F1-value of the BERTScore metric for text similarity [33] and its Huggingface implementation (https://huggingface.co/spaces/evaluate-metric/bertscore). The metric's default model for the English language is "roberta-large". Comparing the "PaGAN sentences" with the concatenated GPT-4 sentences results in BERTScore-F1 = 0.93. The comparison with the GPT-4-generated summary delivers a slightly lower BERTScore-F1 = 0.88. Both values indicate a high similarity between the PaGAN annotations and the GPT-4 results.

Table 1. Results based on 100 patent texts **with** PaGAN contradictions

Metric	
Patents with contradictions	100/100
Number of contradictions found	101
Number of patents with: all PaGAN sentences extracted	36/100
with 1 PaGAN sentence not extracted	37/100
with 2 PaGAN sentences not extracted	17/100
with >2 PaGAN sentences not extracted	10/100

(continued)

Table 1. (*continued*)

Metric	
with all PaGAN sentences <u>not</u> extracted	2/100
Mean number of sentences given in PaGAN dataset per patent	3.81
Mean number of sentences extracted per patent	5.88
Thereof: correctly extracted (true positive)	2.69
Thereof: wrongly extracted (false positive)	3.19
Mean number of sentences not found per patent (false negative)	1.12
Precision	0.46
Recall	0.71
F1	0.56
BERTScore-F1 for concatenated sentences vs. PaGAN sentences	0.93
BERTScore-F1 for summary vs. PaGAN sentences	0.88

Out of the 1,600 patents annotated in the PaGAN dataset as patents **without** contradictions, an arbitrary selection of 100 patents was fed into the LLM *gpt-4-1106-preview* using *LangChain*.

Table 2. Results based on 100 patent texts **without** PaGAN contradictions

Metric	
Patents with contradictions	86/100
Number of contradictions found	87
Mean number of sentences given in PaGAN dataset	0.00
Mean number of sentences extracted	4.84

Table 2 shows that GPT-4 extracted contradictions from 86 of 100 patents, though the human PaGAN annotators tagged all 100 patents as "contradiction-free". For one of the 86 patents even two contradictions were found by GPT-4. With a mean of 4.84, the number of extracted sentences per patent is slightly lower than in the patents with PaGAN contradictions (5.88). One possible explanation for the discrepancy to the PaGAN annotations is that GPT-4 is not good at distinguishing between patents with and without contradictions. An alternative explanation is that the annotations in the PaGAN dataset are not correct, i.e. that "contradiction-free" patents in fact do contain contradictions. A decision between both possible explanations is not finally taken here, but from our perspective the following example of the patent "Golf ball utilizing silicon materials" (US07384349) indicates that at least some of the supposedly "contradiction-free" patents are not annotated correctly in the PaGAN dataset. We find the contradiction extracted by

GPT-4, its summary, parameters ("resilience or carrying distance", "playability characteristics", "multi-layer covers") and the generic TRIZ engineering parameters assigned
("15 - Durability of Moving Object", "33 - Convenience of Use") convincing.

```
{
    "Contradictions": [
        {
            "is_contradiction": "True",
            "target_sentences": [
                "The present invention is directed to new golf ball com-
                positions, preferably utilized in conjunction with mul-
                ti-layer covers, which provide for enhanced coefficient
                of restitution i.e, enhanced resilience or carrying dis-
                tance and/or durability properties when compared to the
                balls found in the prior art.",
                "Despite the great numbers of different materials and
                combinations of materials utilized in prior art golf
                balls, there still remains a need for an improved golf
                ball exhibiting superior properties."
            ],
            "drawback_sentences": [
                "As such, the playability characteristics i.e., feel,
                click, spin, are not diminished."
            ],
            "summary": "Enhanced resilience or carrying distance and
            durability properties are needed without diminishing play-
            ability characteristics such as feel, click, and spin.",
            "evaluation_parameter_1": "resilience or carrying dis-
tance",
            "evaluation_parameter_2": "playability characteristics",
            "triz_parameter_1": "'15 - Durability of Moving Object'",
            "triz_parameter_2": "'33 - Convenience of Use'",
            "action_parameter": "multi-layer covers"
        }
    ]
}
```

5   Conclusion and Future Work

This paper showed that GPT-4, as an example of state-of-the-art LLMs, shows promising
abilities to extract technical contradictions from patents. Compared to the state-of-the-
art of inventive concepts extraction, namely sentence classification with finetuned BERT
derivatives and complicated multi-stage approaches based on the same, our approach can
easily integrate multiple extraction steps in a single prompt and works "out-of-the-box"
with no need for finetuning. The results indicate that a zero-shot *gpt-4-1106-preview* is
at least on par with the best performing fine-tuned sentence classifiers for contradiction
extraction. Despite of its simplicity, the LLM-approach is not limited to the extraction
of sentences, but can also summarize technical contradictions, identify their parameters,
and assign them to generic TRIZ engineering parameters. All these concepts can form
the basis for an improved patent search.

An intended use of the automatically extracted contradictions is to store them in a dedicated retrieval index together with references to their source patents. This allows a cross-domain search for patents addressing contradictions similar to a contradiction currently faced in the innovation process. The solutions from these patents may be transferable to the engineer's own domain resulting in innovative problem solutions.

Weaknesses of the presented approach are a possible lack of reproducibility and the dependency on a commercial black box LLM with comparatively high costs when it comes to the processing of big data like patents. However, the patents only need to be analyzed by an LLM once, during their ingestion into the vector database. At search time, the use of an LLM is not required. Future work can focus on the usage of open-source LLM alternatives to GPT-4 and on further result improvement through Prompt Engineering.

Apart from a case study, our quantitative evaluation was limited to the only publicly available contradiction-annotated dataset "PaGAN". A broader evaluation involving human experts and more expert-annotated data in the validation and refinement of the AI-generated results is desirable.

Furthermore, an extension to more "contradiction-related features", like the assignment of patents to one (or more) of the 40 TRIZ inventive principles is easy to imagine. Finally, the portability of our approach from patents to other document types like scientific papers may be an interesting research topic.

References

1. Spath, D., Warschat, J.: Innovation durch neue Technologien. In: Bullinger, H.-J. (ed.) Fokus Technologie, pp. 1–12. Hanser, München (2008)
2. Slaby, S.: Online Recherche in Patentdatenbanken. Österreichisches Patentamt, Vortragsfolien, Linz (2005). https://silo.tips/downloadFile/online-recherche-in-patentdatenbanken-dr-susanna-slaby. Accessed 03 Mar 2024
3. Altschuller, G.S.: Erfinden – Wege zur Lösung technischer Probleme, 2nd edn. Verlag Technik, Berlin (1998)
4. Zanni-Merk, C.; Cavallucci, D.; Rousselot, F.: Use of formal ontologies as a foundation for inventive design studies. In: Computers in Industry **62**(3), 2011, p. 323–336. https://linkinghub.elsevier.com/retrieve/pii/S0166361510001351. Accessed 03 Mar 2024
5. Giordano, V., Puccetti, G., Chiarello, F., Pavanello, T., Fantoni, G.: Unveiling the inventive process from patents by extracting problems, solutions and advantages with natural language processing. Expert Syst. Appl. **229**(Part A) (2023). https://doi.org/10.1016/j.eswa.2023.120499. Accessed 03 Mar 2024
6. Devlin, J., Chang, M.-W., Lee, K., Toutanova, K.: BERT: pre-training of deep bidirectional transformers for language understanding (2018). https://arxiv.org/abs/1810.04805v1. Accessed 03 Mar 2024
7. OpenAI: GPT-4. https://openai.com/gpt-4. Accessed 03 Mar 2024
8. Srebrovic, R., Yonamine, J.: Leveraging the BERT algorithm for Patents with TensorFlow and BigQuery (2020). https://services.google.com/fh/files/blogs/bert_for_patents_white_paper.pdf. Accessed 03 Mar 2024
9. Guarino, G., Samet, A., Nafi, A., Cavallucci, D.: SummaTRIZ: summarization networks for mining patent contradiction. In: 2020 19th IEEE International Conference on Machine Learning and Applications (ICMLA), pp. 979–986 (2020). https://ieeexplore.ieee.org/document/9356295/. Accessed 03 Mar 2024

10. Orloff, M.A.: Grundlagen der klassischen TRIZ: Ein praktisches Lehrbuch des erfinderischen Denkens für Ingenieure, 3rd edn. Springer, Berlin (2006)
11. Souili, A.; Cavallucci, D.; Rousselot, F.; Zanni, C.: Starting from patents to find inputs to the Problem Graph model of IDM-TRIZ. TRIZ Future 2011, Dublin, Ireland. https://www.resear chgate.net/profile/Achille-Souili/publication/233380622_Starting_from_patent_to_find_i nputs_to_the_Problem_Graph_model_of_IDM-TRIZ/links/5583ee8f08aefa35fe3102c2/Sta rting-from-patent-to-find-inputs-to-the-Problem-Graph-model-of-IDM-TRIZ.pdf. Accessed 03 Mar 2024
12. Souili, A., Cavallucci, D., Rousselot, F.: A lexico-syntactic pattern matching method to extract IDM-TRIZ knowledge from on-line patent databases. TRIZ Future 2012, Lisbon, Portugal. https://www.researchgate.net/publication/275715873_A_lexico-syntactic_Pattern_Matching_Method_to_Extract_Idm-_Triz_Knowledge_from_On-line_Patent_Dat abases. Accessed 03 Mar 2024
13. Heller, L., Warschat, J.: Extraktion von Problemstellung und Lösung aus Patenten mit neuronalen Netzen. In: Bauer, W.; Warschat, J. (eds.) Smart Innovation durch Natural Language Processing, pp. 195–218. Hanser, München (2021)
14. BERT for Patents Model. https://huggingface.co/anferico/bert-for-patents. Accessed 03 Mar 2024
15. Guarino, G., Samet, A., Nafi, A., Cavallucci, D.: PaGAN: generative adversarial network for patent understanding. In: 2021 IEEE International Conference on Data Mining (ICDM) (2021). https://doi.org/10.1109/ICDM51629.2021.00126. Accessed 03 Mar 2024
16. Guarino, G., Samet, A., Cavallucci, D.: PaTRIZ: a framework for mining TRIZ contradictions in patents. Expert Syst. Appl. **207**, 117942 (2022). https://doi.org/10.1016/j.eswa.2022.117942. Accessed 03 Mar 2024
17. Guarino, G.: Text mining for automating TRIZ-based inventive design process using patent documents. Dissertation, University of Strasburg (2022)
18. Geng, R., Li, B., Li, Y., Zhu, X., Jian, P., Sun, J.: Induction networks for few-shot text classification. In: Proceedings of the 2019 Conference on Empirical Methods in Natural Language Processing and the 9th International Joint Conference on Natural Language Processing (EMNLPIJCNLP), pp. 3902–3911 (2019). https://www.aclweb.org/anthology/D19-1403. Accessed 03 Mar 2024
19. Gupta, A.; Thadani, K.; O'Hare, N.: Effective few-shot classification with transfer learning. In: Proceedings of the 28th International Conference on Computational Linguistics, pp. 1061–1066 (2020). https://www.aclweb.org/anthology/2020.coling-main.92. Accessed 03 Mar 2024
20. Kaplan, J., et al.: Scaling laws for neural language models (2020). https://arxiv.org/abs/2001.08361v1. Accessed 03 Mar 2024
21. Bowman, S.R.: Eight things to know about large language models (2023). https://arxiv.org/abs/2304.00612v1. Accessed 03 Mar 2024
22. Gao, T.: Prompting: better ways of using language models for NLP tasks. The Gradient (2021). https://thegradient.pub/prompting/. Accessed 03 Mar 2024
23. Bastian, M.: GPT-4 has a trillion parameters – report. The decoder (2023). https://the-dec oder.com/gpt-4-has-a-trillion-parameters/. Accessed 03 Mar 2024
24. Seals, S.M., Shalin, V.L.: Evaluating the deductive competence of large language models (2023). https://arxiv.org/pdf/2309.05452.pdf. Accessed 03 Mar 2024
25. Zhu, Y., et al.: Large language models for information retrieval: a survey. https://arxiv.org/abs/2308.07107v2. Accessed 03 Mar 2024
26. Brown, T.B., Mann, B., Ryder, N., et al.: Language models are few-shot learners (2020). https://arxiv.org/abs/2005.14165v4. Accessed 03 Mar 2024
27. Reynolds, L.; McDonell, K.: Prompt programming for large language models: beyond the few-shot paradigm (2021). https://arxiv.org/abs/2102.07350v1. Accessed 03 Mar 2024

28. nlpTRIZ: PaGAN Dataset. https://github.com/nlpTRIZ/PaGAN. Accessed 03 Mar 2024
29. LangChain: LangChain. https://www.langchain.com/langchain. Accessed 03 Mar 2024
30. Google LLC: Google Colab. https://colab.research.google.com. Accessed 03 Mar 2024
31. Liu, X., Yu, H., Zhang, H., Xu, Y., et al.: AgentBench: evaluating LLMs as agents (2023). https://arxiv.org/abs/2308.03688v2. Accessed 03 Mar 2024
32. AlpacaEval Leaderboard. https://tatsu-lab.github.io/alpaca_eval/. Accessed 03 Mar 2024
33. Zhang, T., Kishore, V., Wu, F., Weinberger, K.Q., Artzi, Y.: BERTScore: evaluating text generation with BERT (2020). https://arxiv.org/abs/1904.09675. Accessed 03 Mar 2024

AI Based Search Engine to Deploy a TRIZ Pointer to Chemical Effects

Davide Russo[✉] ⓘ, Matteo Cattaneo ⓘ, and Simone Avogadri ⓘ

University of Bergamo, Viale Marconi 5, 24044 Dalmine, Italy
{davide.russo,matteo.cattaneo}@unibg.it

Abstract. Pointers to effects are a group of TRIZ tools which helps the inventor to master a greater knowledge of scientific phenomena and laws, so to suggest him or her different directions to reach possibilities of solution. Pointers to geometric, chemical, and technological effects have been theorized, but only those to physical effects have ever had concrete developments at the research level and as commercial applications.

The aim of this work is twofold: on the one hand, to bring the pointer back to chemical effects (CE), recovering little-known texts that are difficult to find but also difficult to interpret, as they have never been translated from Russian. The other aim is to contextualize these tools in the light of the recent achievements of artificial intelligence technologies in the field of information retrieval. A combination of AI tools, as NER (Named Entity Recognition), RAG (Retrieval Augmented Generation) and LLM (Large Language Model) have been combined in order to identify chemical features from several chemical sources, to index documents in order to answer user's questions, to interact with this Knowledge-Base by a chatbot and finally to generate a complete and standardized output.

A comparison is presented between recent commercial applications of AI and traditional pointers to CE from TRIZ literature. In this paper it is explained how the system works, which are the potentialities according to the AI technologies evolution and a comparative study between a SW infrastructure developed by the authors in collaboration with university spin-off software house and others current AI commercial players like GPT or Gemini based applications.

Keywords: Pointer to chemical effects · AI - Artificial Intelligence · TRIZ · Patents

1 Introduction

Pointers to effect have been theorized for the first time in the 70's by the author of TRIZ Altshuller G.S. and then they have been developed for many years by his students and colleagues. In general this TRIZ tool was one of those with less popularity, except for the pointer to physical effects that has been developed in multiple version. In TRIZ literature it's possible to find different types of pointers to effects: to physical effects, chemical, biological, mathematical, geometrical, engineering and psychological effects [1]. The

© IFIP International Federation for Information Processing 2025
Published by Springer Nature Switzerland AG 2025
D. Cavallucci et al. (Eds.): TFC 2024, IFIP AICT 735, pp. 20–31, 2025.
https://doi.org/10.1007/978-3-031-75919-2_2

aim of these instruments is always to help the designer to find right effects to execute the required function in the system. The most effective way to use these types of tools is to implement them in a functional search logic, as theorised by Litvin in Function Oriented Search (FOS) [2]. For years there have been dozens of statistic applications in which it was mandatory to select a verb from a predefined list as input. With the advent of semantic engines and the possibility of using them on patent databases, this limitation was overcome by giving the user the possibility of using any type of action as input [3].

In a pointer to effects, functions and effects are connected by a library of possible relations. The output can be presented in different ways: a text illustrating the phenomena to be implemented, images or exemplary videos of general solutions but also references to scientific literature or patents (as in Tech-Optimizer and Goldfire Innovators). In the literature, of the types of pointers to effects mentioned above, the most developed have been the physical ones. For the chemical and geometric ones, there are very few dedicated publications, rarely in English [4]. The remaining types of pointers have very few dedicated publications.

The core of the paper is to assess the actuality of this concept of inventive tool and elaborate the proposal of a modern pointer to chemical effects that leverages AI. In following chapter is given a summary of existing main versions of all the pointers including an extended review on existing versions of the chemical pointers. Next chapter present a guideline for creating a chemical pointer to effects based on AI. Section 4 is dedicated to the case study and to explain his results. In the last section conclusions are drawn and new perspectives for future development are given.

2   State of the Art of Pointers to Chemical Effects

The pointer to physical effects was the first type of effects' database to be developed depending of quantity, in fact Savransky [5] affirms that in the 1980's frequency of application of the different kind of effects is 0,2 for geometric, 0,8 for chemical, and 9 for physical (excluding examination of patents in "biology-reach" classes).

With regards to the pointers to chemical effects, it should be noted that chemistry is used as a fundamental source of solutions to TRIZ problems in many publications but works entirely dedicated to this tool are few. In this chapter are presented the main publications about chemical reactions and in general "micro-level" interactions employed as solutions to inventive problems. At this level the concept of chemical contradiction was applied to opposite requirements imposed on substances bound at the molecular level in recent literature [6]. To solve inventive problems at this level are involved micro-objects like: molecule, ion, chemical bond, polarity, etc.

The number of possible chemical effects, according to the perceptions of Markosov, [6], significantly exceeds the number of known chemical compounds, which numbered 50 million [7]. Even if we accept that "only" 350,000 chemical compounds and mixtures of chemicals are registered in the world for production and use, as specified by Wang et al. [8], the number of chemical effects is still so large that scientists, engineers, designers, and TRIZ specialists often do not know how to put them all into practice.

The first remarkable pointer to CE dates to 1988 [9] with Salamatov's work, designed to be used through the table of chemical effects, where the author reports a lot of possible connections between functions and paragraphs, in which is presented at least one chemical substances or process for that function. The library does not cover a great amount of the available literature about chemistry since the pointer is only 66 pages long. Answers from this source are very heterogeneous, as in the majority of other successive pointers: sometimes there is a patent number, at times a patent with relative chemical reaction, and sometimes only the discursive explanation with denomination of chemical substances.

The other most important pointer to CE was built by Mikhailov, implemented in an online library (www.dace.ru*)* [10]. In this database (available only in Russian language) the author organize lots of the available literature about CE in a library giving three possibilities for search strategy: by filling in keywords, by the type of chemical effect selected from a list or by the International Code of Invention (old classification similar to nowadays IPC, International Patent Classification). The output here is a miscellaneous of patent number, textual explanations, chemical reactions and literature references to recover more on the solution proposed (Fig. 1).

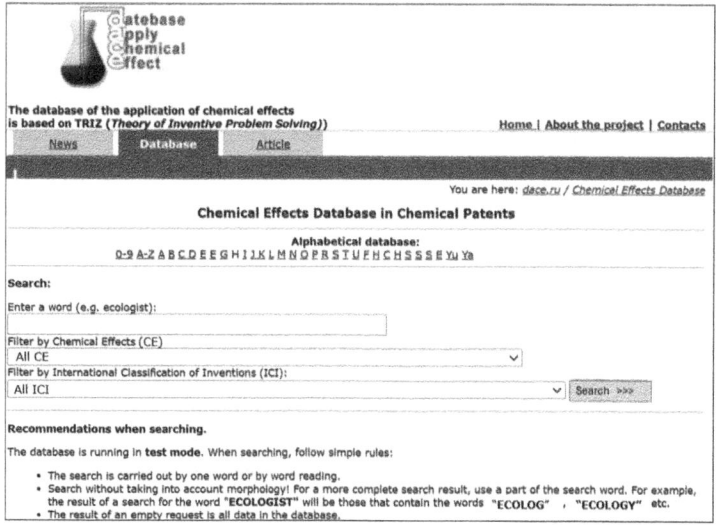

Fig. 1. Translated in English rendering of search interface at dace.ru (available only in Russian)

This main work is the result of a research activity conducted through years. Mikhailov in 1998) [11] published a book in which he focused on the use of chemical effects to solve ecological problems. A long table of effects indexes technical functions, chemical effects, the solving methods of the technical/physical contradiction, the same required reaction and a referment in literature or paragraph to find deeper explanations. In fact, the publication is more in the nature of a manual to the TRIZ methodology applied to the use of pointers to effects.

After that, in 2007 [12] the same author extended his work on CE with another publication that starts with a list of 103 chemical problems explained with their contour conditions. In the second part a small amount of them is analyzed using some of classical

TRIZ tools and in the third part a series of checked solutions is proposed. This is more a collection of case studies than an organized pointer to effects.

One of his last contributions, dated 2012 [13], is a manual about general TRIZ, where the author dedicates a chapter to chemical effects, summarizing in about 40 pages the idea of how should such a pointer work. There are some examples of solved chemical problems but this is considered a very small library of CE, included in this publication to complete an homogeneous state of art on pointer to effects at that time.

Several other sources about this topic were identified in literature, written mainly in Russian, that are not reported due to space constraints [14–18].

It is also interesting to notice how already in 1989 Tsurikov, Litvin and Gerasimov started working on a machine based on first concepts of "Artificial Intelligence" in Minsk founding the company "Invention Machine Corporation". There are publications on the use of this in the field of chemical effects [19].

A common characteristics of all these publications is the subdivision of chemical effects into paragraph or sections. Given the analogical format for the majority of these operas, authors set a classification of effects and the user is so obliged to search the required effect following the logic of preset filters. That means that, if effects are classified for their chemical similarity, is needed to have sufficient knowledges in chemistry to understand where to search for those results. Due to the static nature of the information, it is complicated to find useful results unless they appear explicitly in some table or index provided with the author's ontology. They are also by definition not updatable and as in the case of geometric effects very limited in number and document coverage.

3   How to Build a Pointer to Chemical Effects Based on AI

In these chapter are defined the main steps to realize a modern pointer to chemical effects based on AI tools. In particular the power of these tools is that the quality of answers depends on two basic factors: quality of literature inserted as knowledge-base and the way the AI queries it to process this knowledge and present the exact information requested by the user.

3.1   The Chemical Databases and Books

To create a good knowledge base for this powerful tool, are listed a series of chemical libraries that could be a basic amount of all general chemistry. However, if the problem solving activity is focused on a specific field it may be necessary to add specialistic books (e.g. in the next chapter, "Handbook of petrochemical processes" [20]) since the question is known to be strictly correlated to that topic) (Table 1).

Another fundamental knowledge base for chemical effects are for sure the patents libraries. According to Espacenet (database of the European Patent Office) [21] world families patents in the field of chemistry are more than 9.300.000. Here a table that presents CPC (Cooperative Patent Classification), this helps to understand the variety of information contained in this libraries (Table 2).

Table 1. Available online sources of information for chemical knowledge and databases

NIST Chemistry WebBook	https://webbook.nist.gov/chemistry/index.html.en-us.en
MatWeb	https://matweb.com/
Targetmol	https://www.targetmol.com/index
ChemSpider	https://www.chemspider.com/Default.aspx
PubChem	https://pubchem.ncbi.nlm.nih.gov/classification/#hid=72
Ecoportal	https://ecoportal.su/wastet.html
Academic.ru	https://dic.academic.ru/
Chemical p. of Wikipedia	https://en.wikipedia.org/wiki/Portal:Chemistry
The Merck Index	https://merckindex.rsc.org/
SCIfinder CAS	https://www.cas.org/

Table 2. CPC main categories of class "C" Chemistry (according to European Patent Office)

C01- inorganic chemistry	C11- animal or vegetable oils, fats, fatty
C02- treatment of water, waste water, sewage,	substances or waxes; fatty acids therefrom;
or sludge	detergents; candles
C03- glass; mineral or slag wool	C12- biochemistry; beer; spirits; wine;
C04- cements; concrete; artificial stone;	vinegar;
ceramics; refractories	microbiology; enzymology; […]
C05- fertilizers; manufacture thereof	C13- sugar industry
C06- explosives; matches	C14- skins; hides; pelts; leather
C07- organic chemistry	C21- metallurgy of iron
C08- organic macromolecular compounds;	C22- metallurgy; ferrous or non-ferrous
their preparation or chemical working-up; […]	alloys; treatment of alloys or non-ferrous
C09- dyes; paints; polishes; natural resins;	metals
adhesives; […]	C23- coating metallic material; coating
C10- petroleum, gas or coke industries;	material with metallic material; chemical
technical gases containing carbon monoxide;	surface treatment; […]
fuels; lubricants; peat	C25- electrolytic or electrophoretic processes;
	apparatus therefor
	C30- crystal growth
	C40- combinatorial technology
	C99- subject matter not otherwise provided
	for in this section

3.2 Models of Artificial Intelligence Behind the Inventive Tool

One of the fastest growing sectors right now is AI technologies for text processing. Technologies like LLMs, such as ChatGPT, Gemini, Llama, Claude, and other software applications derived from these same models are replacing the more traditional search engines. The biggest limitation they have is the risk of hallucinations [22].

The authors have developed a model in collaboration with IT experts at Warrant Hub/Trix to adapt a RAG-based chatbot, called Discovery Omnia, to work as a chemical effects pointer. It uses a combination of ChatGPT 4, Meta Llama and Google Gemini 1.5 Pro (compare 4.1). The proposed test saw a comparison between this totally customizable software and other applications on the market. Between these the last version of GPT-4o had been tested: is possible to ask the chatbot for question only relating that file PDF and receive answers with precise references to the text. There are also specific applications that are based on the best-known LLMs such as Perplexity AI: this chatbot permits the user to know which sources are used to give certain answers. It is possible to upload one PDF file per time, the chat must be forced by prompting to recover information from the uploaded material. Some AI pdf-readers were also tested: this online source is only designed to analyze pdf files. Chatbot gives also references to the single page where should be notable information but it's unable to explain which are relevant words. Experimental version of NotebookLM: it analyzes at the same time multiple sources of information (tested with three book at the same time).

3.3 Ideal Output of a Chemical Pointer

To take full advantage of AI features, there are some features that every next-generation pointer should have:

1. The question written by the user should be exploded into a larger description of the problem. An embedded prompt, reformulates the question with words most used in the context being questioned.
2. Related questions. From the initial question, the system can implement a prompt re-routing to understand the type of question and suggesting the resource scheme.
3. Further questions can be added to the initial question to complete the overview of the situation, e.g. suggest in-depth questions about the related industrial process, the limits of feasibility, pros and cons, etc.
4. The answer must allow fact checking, i.e. contain the reference source, with a summary that explains why this source is consistent with the question.
5. The possibility of comparing answers based on documents proposed by the RAG and the more general sources taken from direct training on the Web.
6. Must be as free from hallucinations as possible.
7. Necessary to repost the answer in a simple and understandable way for a non-expert in that specific sector.

4   Case Study: Comparison Between AI Based Chemical Pointers to Effect

The authors want to give the demonstration of how an updated pointer to chemical effects could work effectively with the help of Artificial Intelligence algorithms.

A classical TRIZ case study is how to protect an airplane from explosion when this is hit by a bullet. In this problem one of possible solutions can come from the same fuel of the airplane: it should be liquid in order to burn properly inside the engine but at the same time *it would be so much safer if in solid or gel-like conditions in the storage tanks.*

Once this way of solution is chosen, TRIZ propose to use a pointer to effects to find a working principle to implement this idea. To improve the effectiveness of these tools, a chatbot prompt was developed and an experimental indicator of chemical effects based on patent databases was presented.

4.1 Discovery-Omnia, a RAG Based Pointer on Chemical Patents

This software bases his knowledge on patents from 2020 to 2024 included, of the majority of world's countries. In addition it's feeded with "Horizon Europe" research projects of the same time frame. This information contains the majority of previous technologies in backgrounds. Patents are divided into different sections and the tool search data only in the part, where usually those ones are mentioned (Table 3).

Table 3. Discovery-Omnia for Chemical problems: case study about "how to make a gelled fuel."

User Question
Q1: "Where is used a gelled fuel?"

Discovery-Omnia answers and related suggested patents:

Q* (automatically translated in): *In what applications is gelled fuel used?*
A (From patents and Eu Projects): Let's break down the applications of gelled fuel based on the provided context. The gelled fuel is used in various applications, including: portable flamethrower US20210278177, rocket propellant DE102021124815, portable heater CA3092119, oxyfuel cutting and salting, portable stoves, military applications. Gelled fuel" typically maintains a stable gel state during storage and transportation, and can be transformed into a low viscosity liquid state after being subjected to external shearing during use. CN13461657 Low-molecular mass organic gellant
Concepts:
• Gelled Fuel:A fuel with a stable gel state during storage and transportation, becoming a low-viscosity liquid when sheared during use
• Energetic Metal Nanoparticles (Aluminum, Boron): additives to gelled fuel that increase energy density
• Hydrocarbon Polymers (e.g., Polyethylene Waxes, Long-Chain Fatty Acids, Long-Chain Alcohols):Examples of gelling agents used to solidify oils
• Chromium (III) Acetate:An inorganic crosslinker used with partially hydrolyzed polyacrylamide (PHPAM) to create gelled systems in the oil industry
• Alginate:A gelling agent used in nutritional compositions for tube feeding. It gels under acidic conditions, such as those found in the stomach
• Calcium Salt or Magnesium Salt added to alginate-based compositions. These salts solubilize under acidic conditions and react with the alginate to induce gelling

User Question
Q2: "What is the utilization of gelled fuel in flamethrowers?"

Discovery-Omnia answers and related suggested patents:
Q* (automatically translated in): *What is the utilization of gelled fuel in flamethrowers?*
A (From patents and Eu Projects): A gelled fuel flamethrower is a type of flamethrower that uses a gelled fuel instead of a liquid fuel. The burning rate of gelled fuel is slower than that of liquid fuel [..] The viscosity of gelled fuel may increase compared to liquid fuel, making it more difficult to atomize. Improving the combustion rate of gelled fuel is important for its use. Flammable and low-molecular mass organic gellants with a high energy-containing structure are being sought to enhance the combustion performance of gelled fuel. US20210278177
People also ask: What are the advantages and disadvantages of using gelled fuel in flamethrowers?

A difference between a traditional pointer to effects and AI chatbot based instrument is that, the tool consider previous questions and answers to refine the suggestion to increase relevance for that type of problem. This progressive refining is also possible to be done by hand by drilling down a patent result: selecting patent CN113461657 are proposed other information related to this document and another coherent patent to this function is obtained (EP3307736). Using this source is possible to continue the research by moving backward and forward in patent citation inside the state of art of each patent.

The authors want to focus also technical mistakes let by this pointer:

- "Metal nanoparticles" is correct only for aluminum nanoparticles, but not boron ones. Boron is a metalloid, it is neither a metal nor a non-metal.
- "Hydrocarbon polymers" applies only to polyethylene waxes. Neither long-chain fatty acids nor long-chain alcohols are polymers. Moreover, they are not even hydrocarbons.
- Chromium (III) acetate or chromic salt (III) of acetic acid is an organic, but not an inorganic compound

The tool suggest also an application of gel close to the subject. Pending patent (WO2021242432) that exploit a gel to reduce leakage due to a ballistically caused hole in military tanks of liquid fuel and GB2584963 applies gelled fuel in a process.

4.2   Comparison with Other Available Chemical Pointers

In the following section are presented answers on the same topic from two different types of tools that work as a pointer to effects. On one side there are independent pointers that contain already the knowledge which is asked from the user in different ways. To this category belong classical TRIZ pointer in books and AI chatbots.

Independent Pointers to CE
Are now compared three classical TRIZ pointers [9, 10, 13] with Perplexity.ai and GPT-4. This comparison is made on the base of the same question asked to Discovery in the previous paragraph *"How to get a diesel fuel in a gel condition?"*. These tools don't require any source of knowledge in input because they already embed information.

As shown in Table 4, it is possible to notice that in Salamatov's pointer there is no specific answer on the subject. Findings are presented that could help to find more relevant documents. Results are not easy to search.

In Mikhailov's "Basis of systems theory…" there isn't a specific answer on for the question. There are patent with the use of gel that help to deepen the subject.

In dace.ru the results are coherent, report patents and literature referments, there are filters that, with multiple attempts, help to find relevant results.

Using Perplexity.ai no useful results nor patents number or referment to study the subject. The bot does not understand the request and give proposals to prevent diesel fuel in gel conditions.

With ChatGPT results are complete and very coherent. Each step of the process is explained and is possible to ask for more detail. A minus is absence of sources of information.

Literature Readers Pointer to CE
In this section are questioned chatbot that are AI tools to read some source of information and elaborate an answer based on these sources. In particular were inquired PDF reader like aiPDF [23], NotbookLM [24] and ChatGPT with an uploaded document to be

Table 4. Answers and data given by other tools (AI and traditional ones) to the question "how to make a gelled fuel."

	Salamatov Y.P	Basis of systems theory and [...]	Dace.ru	Perplexity.ai	ChatGPT
Methods to interact with the pointer: "How to get a diesel fuel in a gel condition?"	It is not possible to upload the question. We can enter the function-effect table with "liquefying-thickening" and there is the indication to the paragraph where are summarized uses of gel. It was necessary to search "gel" and "thickening" in the database to obtain the paragraph with 6 patents. This part was selected manually between other 4 false results	The same as the Salamatov's book. Results were found in an alphabetical index under "Gel" This is the unique result about gels in the book	Dace provides "Sol – gel transition" effect in a list of C.E. It gives 53 results, but only 2 are relevants. Combining "Sol – gel transition" with keyword "fuel" in the search mask- a unique result, a relevant one. The filter effect "application of a gel-system" gives a unique result, that is not relevant Using only the keyword "gel" in the search mask gives 55 results, only 2 relevant. If searching for "fuel" as keyword 6 results, 1 relevant. Manual text mining is mandatory to find relevant results	As chat bot, this tool permits to ask any question in natural language. The question was "How to get a diesel fuel in a gel-like consistency?"	As chat bot, this tool permits to ask any question in natural language. The question was "How to get a diesel fuel in a gel-like consistency?"
Results	The system suggests a short description of general and uses of gels In specific, it suggests calcium chloride used as a low melting substances, a German patent with sodium hydrogen phosphate, two USA patent with xylitol and salt hydrates. A thermogel with references to deepen the subject is proposed	No answer is a solution to get a gel fuel. Different application of gel used in 6 patents are quoted: pilework holes, zeolite production, hydrogel, ultrasonic field, protein separation, pressure indicator	For both relevant results is presented a short description of the solution, giving referment and patent to study in depth. "Safe type of fuels". It's explained the use polymer microadditives to obtain gel-like fuels. Are provided 2 patents and 2 referments to deepen solutions. Secondary result: oil with a polymeric additive to get it in gel consistency. Literature recall provided	The chat refuse to give such solution since it could be dangerous or likely illegal. Forcing multiple time to give a technical way, results in ways that, in opposition with question, help fuel to be more fluid, so "no-gels agent"	Steps to gel a fuel: Select a gelling agents. Are proposed silica-based, polymeric and metal soaps Prepare the fuel ensuring no contaminants in it Mixing process Testing and adjustment Storage and handling Example of formulation Safety considerations

analyzed. For these test is chosen the Handbook of petrochemicals [20] as source of information, since it reports this phrase inside the text:

> "Among the diversified uses of naphthenate derivatives is the use of aluminum naphthenate derivatives as gelling agents for gasoline flame throwers (napalm). Manganese naphthenate derivatives are well-known oxidation catalysts."

The ways to interact with this pointers are the same.

Q1: In this book are there information about "How to get a diesel fuel in a gel-like consistency?".

Q2: Are there some "gelling agents" in the book?

Q3: Is there this expression in the book? "gelling agents" (Table 5).

Table 5. Comparison between AI chatbots that work as literature readers

Chatbot	Answer
aiPDF	The answer cite the Fischer–Tropsch process in the book but this is not a way to gel a fuel. About "gelling agents" it answers that there isn't any data
NotebookLM	No data is found inside the given document
ChatGPT	The referment to aluminum and manganese naphthenate is found only with the third question. Not so relevant considering that is a mechanism like manual text-mining

Answering to Q1 and Q2 both aiPDF and ChatGPT quoted the book incorrectly, reporting for Q1 reactions not directly related to the problem and giving no answers for Q2. Asking Q3, that correspond to a manual search using shortcut Ctrl + F in a textual document, aiPDF doesn't find the phrase while ChatGPT find it giving page referment. NotebookLM does not give proper results to any Q1, Q2, Q3 Questions. For these reason this type of tools is judged not mature for the research of technical information typical of a pointer to effect

5 Results

From test on other technologies is possible to say that AI document readers are not ready to analyze critically technical sources of information. The designed AI software Discovery in collaboration with TRIX demonstrate that, as anticipated by the intensive use of patents number in traditional first models of Russian pointers to effect, the knowledge contained in patents are an enormous source of technical information.

The tools (traditional ones and AI based) were tested with further chemical problems which, because of space constraints, are not entirely reported in the paper (Table 6).

Table 6. Some exemplary chemical problems from the test campaign

Questions
Which is the correct percentage of polyols in polyurethan?
Which are industrial technologies to dissolve oxygen in sewage?
How to calibrate plasmonic sensors?
Which are necessary gases for feed cultivation without soil?

(*continued*)

Table 6. (*continued*)

Questions
Which gases work as food preservatives?
How can I purify copper?
Which are methods for carbon capture?
What are Archaea for methanogenesis?
Which is the role of carbon black in the mixture of electrical cable wiring?

6   Conclusions and Future Developments

With this case study is possible to conclude that the nowadays tools of Artificial Intelligence could be merged with knowledge of the chemical pointer to effects to make a complete tool for problem solving. As said above, function are very similar to those of more common pointer to physical effects: since a designer is not interested in means but in results, could be more effective to not distinguish between physical, geometrical and chemical effects proposing a tool that contains all these sections of knowledge.

References

1. Litvin, S.S., Petrov, V.M., Lubin, M.S.: Triz body of knowledge. On site of the TDS Triz Developers Summit (2023). https://triz-summit.ru/en/203941/. (in Russian)
2. Litvin, S.S.: New TRIZ-based tool—function-oriented search (FOS). TRIZ J. 1–12 (2005). https://www.metodolog.ru/triz-journal/archives/2005/08/04.pdf
3. Russo, D., Montecchi, T., Caputi, A.: Tech-finder: a dynamic pointer to effects. INNOVATOR **1**(3), 79–87 (2018)
4. Russo, D., Avogadri, S., Spreafico, C.: AI based pointer to geometric effects. In: Cavallucci, D., Livotov, P., Brad, S. (eds.) TFC 2023. IFIP Advances in Information and Communication Technology, vol. 682, pp. 103–114. Springer, Cham (2023). https://doi.org/10.1007/978-3-031-42532-5_8
5. Savransky, S.D.: Engineering of creativity: introduction to TRIZ methodology of inventive problem solving, p. 359. CRC Press, Boca Raton, New York (2000)
6. Markosov, S.A., Volobuev, D.M., Egoyants, P.A.: Chemical contradiction and methods for resolving it I: Triangle diagram. https://www.metodolog.ru/node/1635 and Chemical contradiction and methods for resolving it II: Modification of precursors. https://www.metodolog.ru/node/1651
7. American Chemical Society. 50 Millionth Unique Chemical Substance Recorded In Chemical Abstracts Service Registry (2009). https://www.sciencedaily.com/releases/2009/09/090910 184310.htm
8. Wang, Z., Walker, G.W., Muir, D.C.G., Nagatani-Yoshida, K.: Toward a global understanding of chemical pollution: a first comprehensive analysis of national and regional chemical inventories. Environ. Sci. Technol. **54**(5), 2575–2584 (2020). https://doi.org/10.1021/acs.est.9b06379
9. Salamatov, Y.P.: Heroic deeds at molecular level/ chemistry helps solve difficult inventive problems. In: Selyutsky, A.B. (ed.) A Thread in the Labyrinth, pp. 95–163. Karelia, Petrozavodsk (1988). ISBN 5-7545-0020-3 (in Russian)

10. Mikhailov, V.A., et al.: Non-commercial project of Chuvash State University named after I.N. Ulianov. Cheboksary (2005–2024). http://dace.ru/db.html. Accessed 05 June 2024. (in Russian)
11. Mikhailov, V.A., Aminov, R.B., Voronina, E.P., Sergeev, S.N., Sokolov, A.Yu.: Solutions of inventive ecological problems. With the use of chemical effects and TRIZ intellectual system, Cheboksary (1998). (in Russian)
12. Mikhailov, V.A., Timokhov, V.I.: Euristics 3- Methodological suggestions to the resolution of chemical problems, Cheboksary (2007). (in Russian)
13. Mikhailov, V.A., Andreev, E.D., Zheltov, V.P., Galetov, V.P., Mikhailov, A.L.: Basis of systems theory and of solutions for inventive technical problems, Cheboksary (2012). (in Russian)
14. Mikhailov, V.A.: Chemical effects and 40 inventive receptions system of G.S. Altshuller and after him, TRIZ-Cheboksary, Chuvash State University (2016). (in Russian)
15. Mikhailov, V.A.: The use of physical and chemical effects for improving chemical systems: methodological suggestions. 1st edn., Cheboksary (1985). 2nd edn., 41 p., Celyabinsk, 1986 (in Russian)
16. Mikhailov, V.A., Kuznecova, T.V.: Chemical effects for engineers. Collection TRIZ-fest 2009, Saint-Petersburg, MATRIZ-SPBSPU (2009). (in Russian)
17. Grierson, B., Fraser, I., Morrison, A., Niven, S., Chisholm, G.: 40 Principles – Chemical Illustrations. Internet journal triz (2003). N6
18. Mikhailov, V., Sosnin, E.: The problem of application of chemical effects in engineering. J. TRIZ 1(14), 52 (2005). (in Russian)
19. Devoino, I.G., et al.: Handbook according to work Artificial Intelligence IM-1.5: BD chemical effects. Minsk, 1989 (look at Seljutskij, How to become an heretic, Petrozavodsk, Karelia, pp 287–295 (1991). (in Russian)
20. Speight, J.G.: Handbook of Petrochemical Processes. CRC Press (2019)
21. European Patent Office CPC. https://worldwide.espacenet.com/patent/cpc-browser. Accessed 05 June 2024
22. Emsley, R.: ChatGPT: these are not hallucinations – they're fabrications and falsifications. Schizophr 9, 52 (2023). https://doi.org/10.1038/s41537-023-00379-4
23. aiPDF. https://aipdf.ai/. Accessed 10 June 2024
24. NotebookLM. https://notebooklm.google.com/. Accessed 10 June 2024

Integrating Generative AI with TRIZ for Evolutionary Product Design

Marin Iuga[1] and Stelian Brad[2(✉)]

[1] Intertechnica SRL, Viseut 26B, Borşa, Maramureş, Romania
marin.iuga@intertechnica.com
[2] Technical University of Cluj-Napoca, Memorandumului 28, 400114 Cluj-Napoca, Romania
stelian.brad@staff.utcluj.ro

Abstract. Democratizing, scaling, and automating creativity and innovation are essential for boosting competitiveness, efficiency, and cost-effectiveness in product creation. This capability will be a key differentiator for organizations seeking a competitive edge and a catalyst for ensuring a better future for humanity. We demonstrate how the TRIZ methodology and Generative AI provide a solid foundation for solving product design challenges. We show how technology and human feedback enable the automated discovery of product design improvements. To prove this, we have developed a technological proof-of-concept solution that leverages the strengths of both TRIZ and Generative AI. This solution explores product reviews, identifies product strengths and weaknesses, and generates an initial palette of potential product design problems that can be iteratively refined to drive product improvement decisions. We provide the theoretical foundation by utilizing TRIZ's systematic problem-solving approach and specific examples of generated data, creating a consistent basis for evaluating the tremendous potential of Generative AI and TRIZ in product design and improvement processes.

Keywords: Generative AI · TRIZ · Product Design · Product Improvement Processes · Innovation Democratization · Creativity Automation

1 Introduction

The advent of Generative Artificial Intelligence (GEN AI) marks a significant milestone in human innovation, as it brings the long-held dream of automating and scaling human creativity to life. This groundbreaking technology is set to transform industries where creativity is paramount – from the world of art to the realm of industrial innovation and even manufacturing.

As GEN AI continues to evolve, its impact on one of the industry's most creative and dynamic processes—product design—will be profound. This transformative potential is already recognized across various sectors. GEN AI enables industrial designers to explore a broader range of ideas and product experiences, including previously unimagined concepts, and to develop initial design concepts significantly faster than traditional

© IFIP International Federation for Information Processing 2025
Published by Springer Nature Switzerland AG 2025
D. Cavallucci et al. (Eds.): TFC 2024, IFIP AICT 735, pp. 32–49, 2025.
https://doi.org/10.1007/978-3-031-75919-2_3

methods [1]. It is essential to understand that Generative AI acts as a catalyst, drastically accelerating and optimizing the product design processes – yet it still needs to be used in conjunction with other methods and technologies to realize its full potential. In principle, GEN AI can generate a large palette of design options, yet human input and expertise are mandatory for validating, refining, and ensuring the feasibility of these potential design options.

Furthermore, GEN AI will be used evolutionarily – optimizing the product design iteration by iteration, along with choosing the most promising candidates after each iteration. One of the complementary methodologies to be used with GEN AI is the Theory of Inventive Problem Solving (TRIZ), which provides a systematic and structured approach to problem-solving, enabling the generation of innovative and effective solutions to complex problems.

Defined as an *"empirical, constructive, qualitative, universal methodology for generating ideas and solving problems (…) on the basis of contradiction models and methods to solve them"* [2], TRIZ assumes a four steps general blueprint for understanding and solving problems in a creative manner [3]:

1. Define the problem: Clearly articulate the specific problem to be solved, gathering relevant data and information to ensure a comprehensive understanding of the issue. This step is crucial, as it sets the stage for the entire TRIZ process.
2. Generalize the problem: Take the specific problem and abstract it to a general problem of principle, identifying the underlying patterns and contradictions that drive the issue, and determining the fundamental conflicts that need to be resolved. This involves recognizing the universal principles that underlie the problem.
3. Find a generalized solution: Utilize the comprehensive TRIZ knowledge base and databases (which include the 40 principles and other TRIZ tools). Use these resources to identify models for solving the problem. Apply these principles and tools to generate a range of potential solutions that address the underlying contradictions and conflicts.
4. Adapt the solution: Refine and tailor the generalized solution to fit the specific context. Ensure that the solution is not only effective in addressing the original problem but also practical and feasible to implement.

Researches on extracting data from various sources (e.g., databases, dataspaces) to be used within or with TRIZ, have been done in papers such as [4–9], and others. In this article, we also explore a proof-of-concept demonstrating how GEN AI enhances the execution of TRIZ processes. Particularly for this research is the consideration of a structured analysis based on main parameters of value (MPVs) to conduct the innovation from data delivered by the GEN AI process. We will examine a scenario in which GEN AI analyzes customer reviews to drive product design improvements and inform TRIZ problem formulation.

2   The Proof-of-Concept Scenario

The general scenario for the proof-of-concept will involve several key steps to demonstrate the integration of GEN AI with TRIZ processes for product design improvement. These steps are detailed in Fig. 1.

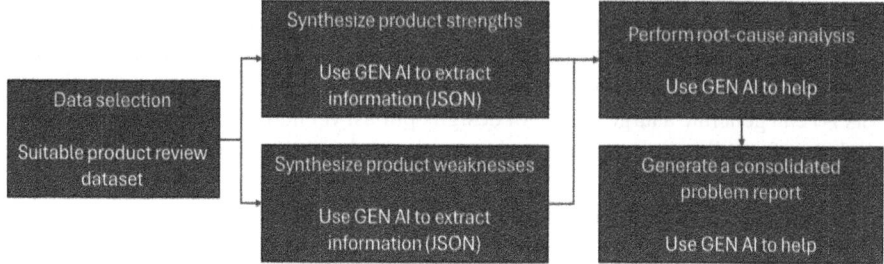

Fig. 1. Data preparation with GEN AI.

1. Data selection: The first step is to select a suitable product review dataset. This dataset should be relevant to the product under analysis and comprehensive enough to provide meaningful insights. The selection process involves ensuring the dataset contains a diverse range of reviews, covering various aspects of the product's performance and user experiences.

2. Synthesizing product strengths: In this step, we use GEN AI to extract information about the product's strengths from the selected dataset. This information is synthesized into a general-purpose data format, with JSON being the preferred format due to its versatility and ease of use. To streamline the process, we focus primarily on positive reviews, which are likely to highlight the product's strengths. The goal is to compile a comprehensive list of positive attributes and features praised by users.

3. Synthesizing product weaknesses: Similarly with the previous step, we extract with the help of GEN AI information about the product's weaknesses from the dataset. This information is also formatted in JSON for consistency and ease of analysis. For simplicity, we focus on neutral and negative reviews, as these are more likely to highlight areas where the product falls short. By identifying common complaints and issues raised by users, we can compile a list of weaknesses that need to be addressed.

4. Performing root-cause analysis: Once the product's weaknesses are identified, we conduct a root-cause analysis with the help of GEN AI to understand the underlying reasons for these issues. This step is crucial as it moves beyond the surface-level symptoms to uncover the fundamental problems affecting the product's design. The outcome of this analysis provides a detailed understanding of the design flaws that need to be corrected, offering essential guidance for future product improvements.

5. Generating a consolidated problem report: After collecting all the data on product weaknesses and the associated design problems, the information needs to be processed and formatted in a manner that is valuable for human consumers. The consolidated problems report will present the findings in an easy-to-understand format, highlighting actionable insights and recommendations. This report will not only serve as a comprehensive overview of the identified issues but also act as a trigger for further product improvements, guiding designers and engineers in their efforts to enhance the product.

By following these steps, the proof-of-concept will demonstrate the effectiveness of combining GEN AI with TRIZ methodologies to analyze customer feedback, identify design issues, and provide actionable insights for product improvement.

2.1   Data Selection

In the proof-of-concept, we will consider an open-source database with product reviews. Specifically, to ensure good data support for GEN AI, we use in this exemplification the Amazon Reviews Dataset focused on musical instruments [10]. The structure of this dataset is the following:

- **reviewerID** - ID of the reviewer, e.g. A2SUAM1J3GNN3B
- **asin** - ID of the product, e.g. 0000013714
- **reviewerName** - name of the reviewer
- **helpful** - helpfulness rating of the review, e.g. 2/3
- **reviewText** - text of the review
- **overall** - rating of the product from 1 to 5
- **summary** - summary of the review
- **unixReviewTime** - time of the review (unix time)
- **reviewTime** - time of the review (raw)

To ensure a relevant selection of data we proceed as follows:

1. Identify the product with the largest number of negative reviews (only reviews that are at least 50 characters long will be considered relevant; shorter reviews will be discarded). This step ensures that we have a sufficient amount of data to accurately determine the product's weaknesses.
2. Merge the **summary** and the **reviewText** content into a single column called **text** which contains the textual content related to the review. We are doing this to simplify the relevant product review content for our analysis.
3. Select only the **text** and the **overall** columns, and retain only the reviews for the product selected in step 1 - thus creating the minimal relevant dataset that will be used for the proof-of-concept.

Fig. 2. D'Addario Accessories Guitar Tuner, a screenshot of a fragment from the Amazon product landing page, taken 10 June 2024.

As a result, the target product processed in this proof-of-concept is the Amazon product with asin B005FKF1PY, more precisely D'Addario Accessories Guitar Tuner (Fig. 2) [11].

For the product in Fig. 2, there are 47 positive reviews (4 or more stars) and 16 neutral or negative reviews (3 or less stars) in the dataset. While for production-relevant use cases, the volume of data should be higher, for the proof-of-concept scenario the data is sufficient (even more, we prefer for now a small amount of data to validate the behavior on small-data datasets).

Having this data, we proceed further with performing cognitive tasks using GEN AI. The LLM model selected for this job is **llama-3-sonar-large-32k-chat** [12]. Llama-3-sonar-large-32k-chat is designed for conversation and excels at chat interactions. This model is open-source and it is likely chosen for its ability to hold fluent and engaging conversations, making it a good fit for this purpose.

2.2   Synthetizing Information on Product Strengths

We use GEN AI to analyze the positive product reviews (4 or more stars) and extract the product strengths. For each product strength, we are interested in a short title, a general description, relevancy and frequency scores (between 1 and 5) along with product review content fragments which are most relevant for each of the product strength.

To make sure we are extracting the data in a manner that can be further refined using GEN AI (or even general-purpose algorithms) we extract the data in a JSON list format. Each JSON item from the list is associated with a product strength and has the following fields:

- **title**: the title of the strength.
- **description**: the description of the strength, we aim for around 3 sentences for description.
- **relevancy score**: an estimate of the relevancy of the strength item as a number between 1 (least relevant) and 5 (most relevant).
- **frequency score**: an estimate of the frequency of the strength item in the product review content. It will be a number between 1 (least frequent) and 5 (most frequent);
- **keys**: the keys identifying the most relevant reviews for the strength. We consider this field as support for further processing.
- **samples**: a list of fragment content samples from reviews which are most representative for the strength.

We further perform the following steps:

1. Transform the positive review data into a JSON collection, each item from the collection being represented by a review key (system generated) and the review text.
2. Create a GEN AI prompt that includes the JSON collection created in the previous step and instructs the model to generate the expected output in the JSON format previously described.
3. Capture the generated output and inspect it for compliance with the expected format and the expected values.
4. If the generated output is not compliant, attempt a new inference session.

5. If the generated output is compliant, save the generated JSON data for further processing.

By following these steps, the GEN AI model has identified for the product in Fig. 2 the following major strengths (both the generated names and descriptions of the product's strengths reproduced as-is):

"*Accurate and Reliable Tuning - The tuner is highly accurate and reliable, providing quick and precise tuning results. Many users have reported that it is as accurate as more expensive tuners, and it has become their go-to tuner for gigs and practice sessions. The tuner's ability to detect vibrations and provide clear visual feedback makes it easy to use, even for beginners.*

Compact and Discreet Design - The tuner's compact and discreet design makes it easy to use on stage or in practice sessions without drawing attention to itself. Many users have reported that it is almost invisible when attached to the headstock, and it doesn't obstruct their playing. The small size also makes it easy to store in a guitar case or bag.

Easy to Use and Intuitive Interface - The tuner's interface is easy to use and intuitive, with a clear and bright display that makes it easy to read. Many users have reported that it is easy to navigate, even for those who are not tech-savvy. The tuner's simplicity and ease of use make it a great option for beginners and experienced musicians alike.

Convenient and Portable - The tuner is highly portable and convenient, making it easy to take to gigs, practice sessions, or anywhere else you need to tune your instrument. Many users have reported that it is easy to attach and detach, and it fits easily in a guitar case or bag.

Good Value for the Price - The tuner is a great value for the price, offering accurate and reliable tuning at an affordable cost. Many users have reported that it is a great option for those on a budget, and it is a good alternative to more expensive tuners."

A JSON sample generated for one of the product strengths is shown in Fig. 3.

{
  "title": "Accurate and Reliable Tuning",
  "description": "The tuner is highly accurate and reliable, providing quick and precise tuning results. Many users have reported that it is as accurate as more expensive tuners, and it has become their go-to tuner for gigs and practice sessions. The tuner's ability to detect vibrations and provide clear visual feedback makes it easy to use, even for beginners.",
  "relevancy_score": 5,
  "frequency_score": 5,
  "keys": "0, 11, 24, 39, 51",
  "samples": [
    "I have used this for a couple of months now, switching from electric to acoustic to classical guitars several times daily and am happy to report it works and works well.",
    "It is very precise and accurate, leading you to perfect tuning every use.",
    "I've compared this little guy to three other chromatic tuners and it's dead accurate.",
    "This tuner will get you close quickly and you can use your ears to fine tune if necessary.",
    "It is real accurate, snaps on to the headstock perfectly."
  ]
}

Fig. 3. A product strength outcome sample synthetized in JSON data format by using GEN AI.

We can observe a high degree of coherence among product strength's title, description and samples. The GEN AI model synthetized valuable and meaningful information, as well as was capable to select original content which is highly relevant to the newly generated information. The main product strengths, along with their impact score (relevancy score multiplied by the frequency score), – as assessed by the GEN AI model - can be visualized in Fig. 4.

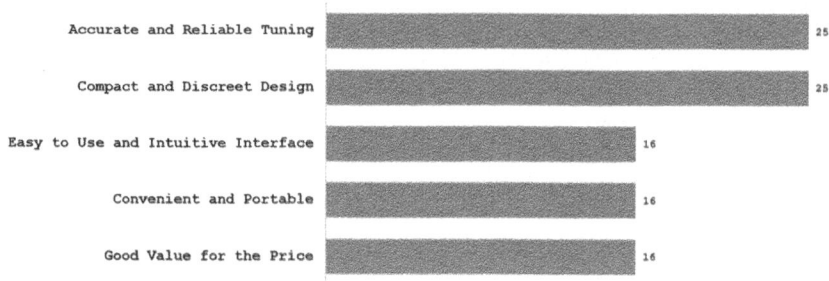

Fig. 4. Most relevant product's strengths and their impact.

At this point, by successfully utilizing GEN AI, we can form a general idea of the product strengths, an important factor in determining the possible range of solutions for the potential weaknesses and challenges that can be detected at a later stage.

2.3 Synthetizing Information on Product Weaknesses

We perform a similar job to synthetize information about the product's weaknesses. The main difference is that the source of initial information is the content from the neutral or negative reviews. Otherwise, the JSON format is like the one used for capturing the product's strengths and the steps used for generating the product's weaknesses information are identical. The GEN AI prompt is fine-tuned to refer to weaknesses instead of strengths. Therefore, the GEN AI model determined the following product's weaknesses (generated names and descriptions being reproduced as-is):

"Difficulty in Reading Display in Bright Light - The tuner's display is difficult to read in bright light conditions, making it challenging to accurately tune instruments. This issue is particularly problematic for outdoor performances or well-lit environments. Users have reported struggling to see the screen, which can lead to frustration and tuning errors.

Battery Life and Drain Issues - The tuner's battery life is short, and some users have reported that it drains quickly, even when turned off. This can be frustrating for users who need reliable performance without frequent battery replacements. The tuner's power management seems to be inefficient, leading to unexpected battery drain.

Challenging User Interface - Some users have found the tuner's interface to be challenging to use, with difficulties in determining button functions and power controls. The small size of the tuner can make it hard to navigate, leading to user frustration and errors.

Limited Tuning Responsiveness - The tuner has limited responsiveness, requiring users to strum strings loudly and position it in specific locations on the headstock for accurate tuning. This can lead to inefficient tuning processes and potential frustration for users.

Fragile Build Quality - Some users have expressed concerns about the tuner's build quality, mentioning issues with loose battery covers, fragile ratcheting clips, and overall durability. This can affect the longevity and reliability of the tuner, potentially leading to premature breakdowns."

A JSON sample for the most representative weakness is shown in Fig. 5.

```
{
  "title": "Limited Tuning Responsiveness",
  "description": "The tuner has limited responsiveness, requiring users to strum strings loudly and position it in specific locations on the headstock for accurate tuning. This can lead to inefficient tuning processes and potential frustration for users.",
  "relevancy_score": 5,
  "frequency_score": 3,
  "keys": "60",
  "samples": [
    "To get it to work one has to place it low on the headpiece (next to the nut) and strum a string pretty loud for a consistent register.",
    "Placing it at the top of the headpiece it simply doesn't pick up enough string vibration to register well.",
    "You can always clip it on the top of the headpiece and then move it when you're ready to tune."
  ]
}
```

Fig. 5. A product weakness outcome sample synthetized in JSON data format with GEN AI.

We can observe again a high coherence between the generated title and description, along with the high relevance of selected samples. The GEN AI model has performed well both in generative and discriminative operation modes. Additionally, the product weaknesses and their impact can be visualized in Fig. 6.

Product weaknesses and their impact

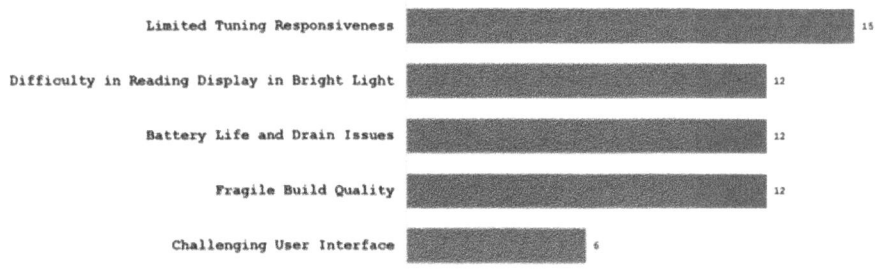

Fig. 6. Most relevant product's weaknesses and their impact.

First, we can observe that the product's weaknesses have a much lesser impact than the product's strengths. This is coherent with the fact that there are roughly three times more positive reviews than neutral or negative reviews. We can also observe that, regarding the product's tuning, there are split opinions: some users consider it accurate and reliable, while other users consider it as having low responsiveness. It is an issue that needs to be followed up since maybe some of tuning characteristics are appealing to some users and not appealing to others. However, the GEN AI model has comprehensively captured the relevant weaknesses of the product.

2.4   Performing Root-Cause Analysis

We have successfully utilized GEN AI to comprehensively analyze the product, thereby identifying both its strengths and weaknesses. By understanding weaknesses, we can identify their root causes and articulate the problems associated with the product design. While understanding the product's strengths is not a primary focus at this stage, it will become crucial when proposing solutions – as we need to ensure that these solutions will not compromise or invalidate the product's existing strengths and benefits (thus, the important role of TRIZ contradictions analysis).

To identify the product design problems using GEN AI, we instruct the model to perform the following actions for each product weakness identified in the previous step:

1. Consider all the negative reviews as the primary data source for identifying the weakness's root causes.
2. Consider the weakness's title, description, and samples as the guiding factor for root cause identification.
3. For each identified root cause, ensure that the following is generated:
   a. The name of the root cause in generic, high-level terms,
   b. The problem associated with the root cause, ensuring that it is specified what exactly the problem is, who is affected by the problem, and how it is affected by the problem.
4. Execute a GEN AI prompt which determines the model to consider all the steps above and return the output in a JSON format strictly controlled by the prompt structure. Ensure that the JSON data links the identified root cause with its associated product design weakness.

It should be noted that the prompt allows the model for describing the problems using also information encoded in the model itself, even if this may lead to model hallucination issues. This was done in order to emulate a brainstorming process where the model itself is an active participant. Further product design iterations may ground the problem formulations (for example using a factual knowledge database, RAG techniques, or even humans in the loop), yet for now we are fine with some creativity. A JSON sample data item generated for the root cause analysis of a specific weakness is shown in Fig. 7.

```
[
  {
    "weakness_name": "Limited Tuning Responsiveness",
    "name": "Inadequate Sensor Design",
    "problem_formulation": "The tuner's sensor is not sensitive enough to detect string vibrations accurately,
    leading to inconsistent tuning results, frustrating users, and wasting their time."
  },
  {
    "weakness_name": "Limited Tuning Responsiveness",
    "name": "Insufficient Power Source",
    "problem_formulation": "The tuner's power source is not sufficient to support consistent and accurate tuning,
    resulting in unreliable performance, and impacting users who rely on it for quick and efficient tuning."
  },
  {
    "weakness_name": "Limited Tuning Responsiveness",
    "name": "Poor Ergonomic Design",
    "problem_formulation": "The tuner's design does not accommodate various headstock shapes and sizes, making it
    difficult for users to position it correctly, and leading to inaccurate tuning results and user frustration."
  }
]
```

Fig. 7. A root analysis outcome sample synthetized in JSON data format by using GEN AI and associated with a specific product weakness.

We observed that the model captures relevant potential problems regarding product design (for other weaknesses it was also able to identify potential production process design issues). We also observed that the problem formulation is highly coherent with the associated weakness and root cause. Empirically, the problem formulations seem to be sound, however, for future iterations of product design, we will possibly need human expertise to correct and guide the model (actually the entire AI ecosystem) further.

2.5 Creating the Product Design Problems Report

At this point, whit a significant leverage of GEN AI capabilities, we have gathered all the relevant data needed to identify and describe the problems associated with the product's design.

We need to present this information in a way that makes it truly valuable: clear, actionable, and impactful, thereby triggering the desired response and driving meaningful outcomes. This is a challenge that can be addressed by standard software engineering practices as we have all the data, we just need to consolidate and format it for human communication. We can still use the GEN AI to further adjust our communication, for example to summarize the data and create a call to action.

The information from the following report (Table 1) regarding product weaknesses and product design problems was created with GEN AI, including the closing remarks section at the end of the report.

Table 1. The report created by GEN AI.

D'ADDARIO ACCESSORIES GUITAR TUNER
POTENTIAL PRODUCT DESIGN PROBLEMS REPORT

WEAKNESS 1: Difficulty in Reading Display in Bright Light
The tuner's display is difficult to read in bright light conditions, making it challenging to accurately tune instruments. This issue is particularly problematic for outdoor performances or well-lit environments. Users have reported struggling to see the screen, which can lead to frustration and tuning errors

PROBLEM	DESCRIPTION
Insufficient Display Brightness	The tuner's display is not bright enough to be readable in well-lit environments, causing frustration and tuning errors for musicians who perform outdoors or in bright settings
Inadequate Screen Technology	The screen technology used in the tuner is not suitable for high-light conditions, making it difficult for musicians to accurately tune their instruments in outdoor or well-lit environments
Poor Display Design	The display design of the tuner is not optimized for readability in bright light, resulting in difficulties for musicians to quickly and accurately tune their instruments in high-light settings

WEAKNESS 2: Battery Life and Drain Issues
The tuner's battery life is short, and some users have reported that it drains quickly, even when turned off. This can be frustrating for users who need reliable performance without frequent battery replacements. The tuner's power management seems to be inefficient, leading to unexpected battery drain

PROBLEM	DESCRIPTION
Inefficient Power Management	The tuner's power management system is not optimized, leading to rapid battery drain, even when the device is turned off, causing frustration and inconvenience for musicians who rely on the tuner for performances and practice sessions
Poor Battery Design	The battery design is flawed, resulting in a short battery life and unexpected drain, making it difficult for musicians to trust the tuner's performance and plan their practice sessions and performances accordingly
Loose Battery Cover	The battery cover is not securely designed, allowing the battery to fall out or become dislodged, leading to unexpected battery drain and inconvenience for musicians who need a reliable tuner

(continued)

Table 1. (*continued*)

WEAKNESS 3: Challenging User Interface	
Some users have found the tuner's interface to be challenging to use, with difficulties in determining button functions and power controls. The small size of the tuner can make it hard to navigate, leading to user frustration and errors	
PROBLEM	DESCRIPTION
Poor Button Design	The tuner's buttons are too small and lack clear indication of their functions, making it difficult for users to navigate and control the device, leading to frustration and errors
Inadequate Display Design	The tuner's display is too small and uses a chunky LCD block style, making it hard for users to read and understand the information, leading to difficulties in tuning and user frustration
Lack of Intuitive Controls	The tuner's controls are not intuitive, requiring users to learn and remember specific button combinations and functions, leading to user confusion and errors

WEAKNESS 4: Limited Tuning Responsiveness	
The tuner has limited responsiveness, requiring users to strum strings loudly and position it in specific locations on the headstock for accurate tuning. This can lead to inefficient tuning processes and potential frustration for users	
PROBLEM	DESCRIPTION
Inadequate Sensor Design	The tuner's sensor is not sensitive enough to detect string vibrations accurately, leading to inconsistent tuning results, frustrating users, and wasting their time
Insufficient Power Source	The tuner's power source is not sufficient to support consistent and accurate tuning, resulting in unreliable performance, and impacting users who rely on it for quick and efficient tuning
Poor Ergonomic Design	The tuner's design does not accommodate various headstock shapes and sizes, making it difficult for users to position it correctly, and leading to inaccurate tuning results and user frustration

WEAKNESS 5: Fragile Build Quality	
Some users have expressed concerns about the tuner's build quality, mentioning issues with loose battery covers, fragile ratcheting clips, and overall durability. This can affect the longevity and reliability of the tuner, potentially leading to premature breakdowns	
PROBLEM	DESCRIPTION
Inadequate Material Selection	The tuner's components, such as the battery cover and ratcheting clip, are made of low-quality materials that are prone to breakage or malfunction, leading to user frustration and premature replacement
Poor Design Engineering	The tuner's design does not adequately consider the stresses and strains of regular use, resulting in a fragile build that is susceptible to damage and failure, impacting users who rely on the tuner for accurate and reliable tuning
Insufficient Quality Control	The manufacturer's quality control processes are inadequate, allowing defective or subpar tuners to reach the market, which can lead to user dissatisfaction and a loss of trust in the brand

CLOSING REMARKS
The D'Addario Accessories Guitar Tuner excels in providing accurate and reliable tuning, compact and discreet design, ease of use, portability, and value for the price. However, it falls short in several areas, including readability in bright light, battery life and drain issues, challenging user interface, limited tuning responsiveness, and fragile build quality. These issues starkly contrast with the tuner's strengths, highlighting the need for targeted improvements. For instance, the tuner's compact design is compromised by the difficulty in reading the display in bright light, and its ease of use is hindered by the challenging user interface. Similarly, the value for the price is diminished by the battery life and drain issues and fragile build quality. The tuner's accuracy and reliability are also undermined by the limited tuning responsiveness and poor ergonomic design
It is crucial to address these product design problems urgently to preserve and augment the strengths of the D'Addario Accessories Guitar Tuner. If left unaddressed, these issues can erode user trust and satisfaction, ultimately impacting sales and the brand's reputation. Furthermore, the tuner's strengths, such as its accuracy and reliability, may be overshadowed by its weaknesses, leading to a loss of market share. By addressing these problems, the brand can not only maintain its competitive edge but also enhance the overall user experience, fostering loyalty and driving growth. The urgency to act is further emphasized by the potential for users to seek alternative products that better meet their needs, leading to a loss of market share and revenue
We strongly recommend allocating resources to finance further product improvements, focusing on the identified weaknesses. By doing so, we can ensure that the D'Addario Accessories Guitar Tuner continues to excel in the market, solidifying its position as a reliable and user-friendly tuning solution. This investment will not only benefit the product but also the brand as a whole, ultimately driving business success and customer satisfaction. We urge decision-makers to take prompt action to address these issues, securing the future of the product and the brand

We can now easily observe the significant extent to which GEN AI has facilitated the successful completion of the Problem Definition stage in TRIZ for product improvement. The data synthesized using GEN AI is meaningful, comprehensive, and very useful. It is clear, actionable, and impactful – as required. Furthermore, the degree of the model's hallucinations is very low, making the information even more valuable.

GEN AI facilitated the creation of the product design problems report, an artifact that would have required a significant investment of human effort and expertise if done manually. Even more, the manual effort would have been needed again and again for every new product – yet by using GEN AI automation - a new report for a new product can be generated very fast and with minimal investment.

3 Product Advancement

3.1 Analysis of Main Parameters of Value and Critical Weaknesses

At this point, Information from Sect. 2 can be used for better formulating the main parameters of value (MPVs) for product improvement. Information from Sect. 2 also gives us the possibility to identify the impact and current satisfaction for each MPV. This is a major support provided by GEN AI. With the help of GEN AI, we can partially automate the extraction of MPVs, as well as analysis of weaknesses. The graph from Fig. 8 plots MPVs with their current satisfaction levels on the x-axis and impact scores on the y-axis (as extracted with the help of GEN AI). Each point represents an MPV, and the color coding corresponds to different MPVs as indicated in the legend.

Key observations from the analysis reveal that "Compact and Discreet Design" and "Accurate and Reliable Tuning" are well-performing MPVs with both high satisfaction and high impact. Conversely, "Difficulty Reading Display" has a relatively high impact but low satisfaction, indicating a critical area needing improvement. Additionally, "Battery Life and Drain Issues," "Limited Tuning Responsiveness," "Fragile Build Quality", and "Challenging User Interface" are moderate to low impact but have low satisfaction, highlighting several areas requiring attention for enhancement. Meanwhile, "Convenient and Portable" and "Good Value for the Price" have moderate impact and high satisfaction, suggesting they are strengths that should be maintained. Actionable insights suggest

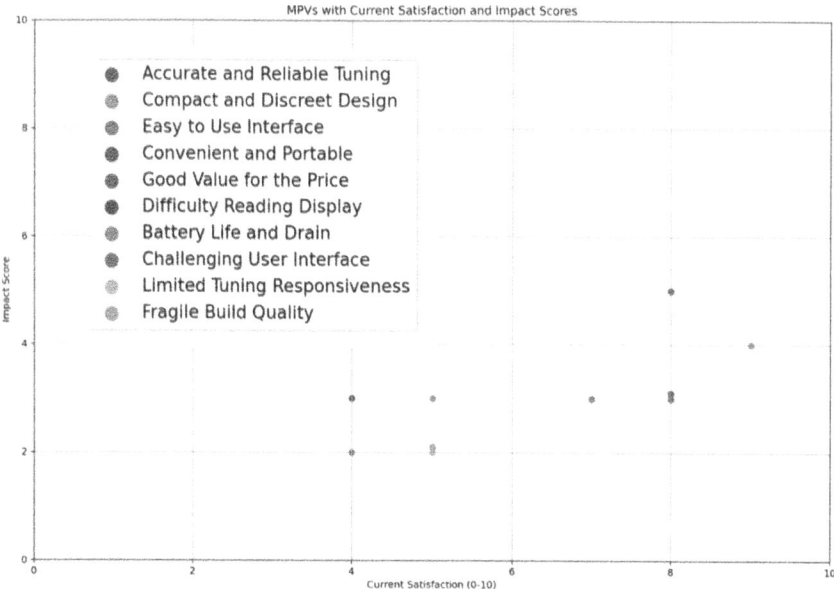

Fig. 8. MPVs and their state in the framework of product competitiveness.

prioritizing improvements in areas with low satisfaction but high impact, such as "Difficulty Reading Display". It is also crucial to maintain strengths like "Compact and Discreet Design" and "Accurate and Reliable Tuning," which are already performing well. Furthermore, focusing on moderate improvements for attributes like "Battery Life and Drain Issues" and "Challenging User Interface" can significantly enhance overall user satisfaction.

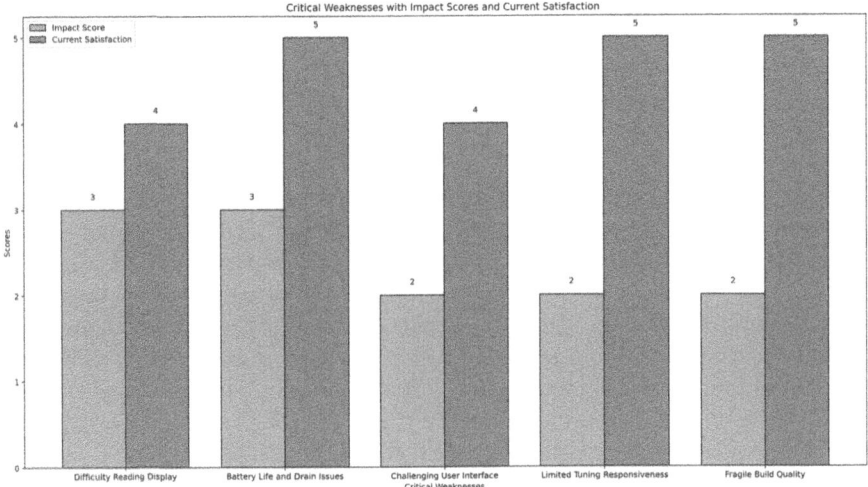

Fig. 9. Critical weaknesses and their status.

Figure 9 highlights the critical weaknesses with their impact scores and current satisfaction levels. The critical weaknesses identified include "Difficulty Reading Display in Bright Light" (Impact Score: 3, Current Satisfaction: 4), which has a moderate impact but relatively low satisfaction, indicating the need for improved display readability in bright light conditions. "Battery Life and Drain Issues" (Impact Score: 3, Current Satisfaction: 5) also show moderate impact with low satisfaction, highlighting dissatisfaction with battery performance and the need for optimization. The "Challenging User Interface" (Impact Score: 2, Current Satisfaction: 4) reflects lower impact but still low satisfaction, suggesting that simplifying and improving the interface can significantly boost user satisfaction. "Limited Tuning Responsiveness" (Impact Score: 2, Current Satisfaction: 5) indicates a lower impact score with similarly low satisfaction, necessitating enhancements in responsiveness. Lastly, "Fragile Build Quality" (Impact Score: 2, Current Satisfaction: 5) reveals concerns about build quality with a lower impact score and low satisfaction, underscoring the need to increase durability. Key takeaways and actions include high-priority improvements such as implementing enhanced display technologies (e.g., anti-glare coating, adaptive brightness) for "Difficulty Reading Display" and optimizing power management and using higher capacity batteries for "Battery Life and Drain Issues". Moderate priority improvements involve redesigning the interface to be more intuitive with larger, clearly labeled buttons and touch screen capabilities for "Challenging User Interface" and upgrading sensors and improving tuning algorithms for better responsiveness in "Limited Tuning Responsiveness". Low-priority but necessary improvements include using more durable materials and reinforcing structural design to enhance durability for "Fragile Build Quality".

3.2 Inventing High-Impact MPVs and Ensuring High Satisfaction

We see in Fig. 8 that no high-impact MPVs have been revealed from market analysis. This is an opportunity for further product innovation. Given the need to introduce MPVs with high impact and to ensure these new MPVs meet high satisfaction, we can utilize the TRIZ methodology to systematically develop innovative solutions. The proposed high-impact MPVs and potential solutions are further illustrated.

1. Adaptive display technology

   - Description: A display that adapts to various lighting conditions, providing excellent readability both indoors and outdoors.
   - TRIZ principle: Segmentation - Use segments of adaptive e-ink or OLED panels that adjust brightness and contrast based on ambient light sensors.
   - Ensuring satisfaction: Implement user control for manual adjustments and provide presets for different environments.

2. Long-lasting battery with quick recharge

   - Description: A battery system that offers extended life with rapid recharging capabilities.
   - TRIZ principle: Dynamics - Introduce a dynamic power management system that intelligently optimizes energy usage based on activity.
   - Ensuring satisfaction: Include fast-charging technology and provide a clear battery status indicator to avoid unexpected shutdowns.

3. Intuitive touchscreen interface

   - Description: A highly intuitive touchscreen interface that simplifies user interactions.
   - TRIZ principle: Feedback - Incorporate haptic feedback and context-sensitive controls to enhance usability.
   - Ensuring satisfaction: Offer customizable interface settings and interactive tutorials to cater to both novice and advanced users.

4. Advanced sensor technology for tuning

   - Description: Highly responsive sensors that provide precise tuning with minimal effort.
   - TRIZ principle: Increasing the degree of system dynamics - Use multi-point detection sensors that adapt to different instruments and playing conditions.
   - Ensuring satisfaction: Ensure that sensors are accurate across various string instruments and provide clear visual/auditory feedback during tuning.

5. Durable and eco-friendly build

- Description: A tuner built with durable, sustainable materials that withstands rigorous use.
- TRIZ principle: Composite materials - Utilize advanced composite materials that are both strong and environmentally friendly.
- Ensuring satisfaction: Emphasize durability and sustainability in marketing to attract eco-conscious consumers and provide a warranty for durability assurance.

3.3 Addressing Critical Weaknesses

Some of the critical weaknesses are addressed by the solutions proposed in Sect. 3.2. To address the remaining critical weaknesses identified in the analysis, we have applied TRIZ principles to develop inventive solutions. This refers to "Limited Tuning Responsiveness" and "Fragile Build Quality". The solutions to these weaknesses will also indirectly improve other areas of concern by enhancing the overall functionality and durability of the product.

For "Limited Tuning Responsiveness", the current issue is characterized by low impact and low satisfaction due to slow and inaccurate tuning response. Using the TRIZ principles of "Nested Doll" and "Self-Service", we propose the following inventive solutions: First, by developing multi-point detection sensors that can pick up vibrations from different parts of the instrument, we enhance accuracy through cross-referencing multiple input points. Second, implementing advanced signal processing algorithms that adapt to different instruments and playing styles will ensure consistent tuning accuracy across various conditions.

For "Fragile Build Quality", the current issue is marked by low impact and low satisfaction due to concerns about durability. Applying the TRIZ principles of "Composite Materials" and "Asymmetry", we suggest using high-strength, lightweight materials such as carbon fiber composites or reinforced plastics to enhance durability without adding weight. Additionally, designing the tuner with an asymmetrical structure provides better stress distribution and reduces points of failure. These improvements tackle the specific weaknesses and contribute to the overall robustness and user satisfaction of the product.

3.4 Conflict Check and Synergy

There are no conflicting issues with the proposed improvements. The adaptive display technology enhances readability without compromising battery life, thanks to the segmented display approach. The long-lasting battery with quick recharge improves battery performance, complementing the adaptive display and advanced sensor technology without conflicting with other features. The intuitive touchscreen interface simplifies user interactions, aligning well with advanced sensor technology and improving overall user experience without adding complexity. The advanced sensor technology for tuning provides precise tuning, working synergistically with the durable build and intuitive interface to enhance accuracy and usability. Additionally, the durable and eco-friendly build enhances durability without adding weight, complementing all other improvements by providing a robust and reliable product foundation.

There are several synergies between the proposed improvements. The combination of adaptive display technology and a long-lasting battery ensures that improved readability does not come at the cost of reduced battery life. The intuitive touchscreen interface and advanced sensor technology work together to provide an enhanced user experience, making tuning more accurate and user-friendly. The durable and eco-friendly build supports all technological advancements by ensuring the product remains reliable and sustainable, thus increasing overall satisfaction.

3.5 Marketing-Driven Presentation

With the help of GEN AI, the characteristics of the new product can be used to elaborate a professional presentation, as well as a virtual prototype. They can be used for fast testing in a promoter light test or focus group, as shown in Table 2.

Table 2. Content for early-stage market testing created with GEN AI.

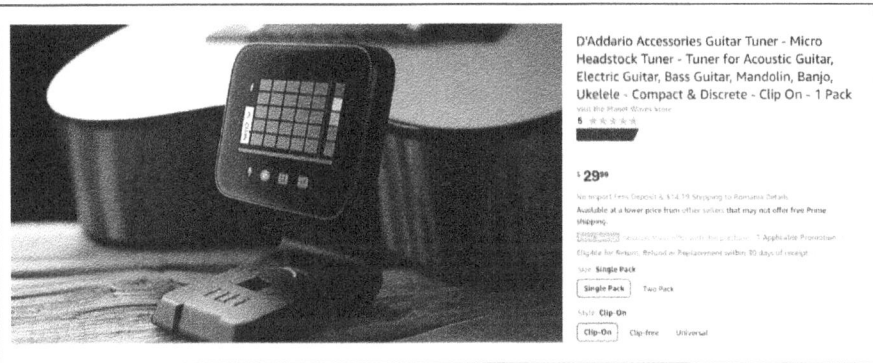

D'Addario Accessories Guitar Tuner - Micro Headstock Tuner - Tuner for Acoustic Guitar, Electric Guitar, Bass Guitar, Mandolin, Banjo, Ukelele - Compact & Discrete - Clip On - 1 Pack

The tuner features an adaptive display that seamlessly transitions between segmented e-ink and OLED panels, ensuring clear readability in all lighting conditions, both indoors and outdoors. Internally, the device houses a dynamic power management system connected to a long-lasting, high-capacity battery that supports rapid recharging. This system intelligently optimizes energy usage based on activity, ensuring extended battery life without compromising performance. A clear battery status indicator on the display helps users monitor power levels to avoid unexpected shutdowns. The intuitive touchscreen interface, made from durable, scratch-resistant glass, provides haptic feedback and context-sensitive controls, making interactions smooth and responsive. The interface is customizable, allowing users to tailor settings to their preferences, and includes interactive tutorials to assist both novice and advanced users. At the core of the tuner are advanced multi-point detection sensors capable of picking up vibrations from various parts of the instrument. These sensors, coupled with sophisticated signal processing algorithms, adapt to different instruments and playing styles, ensuring consistent and precise tuning. Visual and auditory feedback is provided to guide users through the tuning process efficiently. The tuner is encased in a robust shell made from high-strength, lightweight carbon fiber composites and reinforced plastics, ensuring durability without adding unnecessary weight. The asymmetrical internal structure distributes stress evenly, reducing points of failure and enhancing overall reliability. The materials used are eco-friendly, underscoring the product's commitment to sustainability. Additionally, the device comes with a warranty for durability, appealing to eco-conscious consumers.

Determining the right price for the new tuner involves considering several factors, including production costs, perceived value, market competition, and customer willingness to pay. We have material costs: $7, manufacturing costs: $4, packaging and distribution costs: $2, and other costs (R&D, marketing, etc.): $2. This means we have total production costs of $15. Typically, companies aim for a profit margin between 30% and 50%. We consider 40% profit margin. Thus, the cost-based price is $21.

Assessing the significant improvements (adaptive display, long-lasting battery, touchscreen interface, advanced sensors, durable build), we need to determine how much more customers are willing to pay for these enhancements. Focus group is used to estimate this perceived value. In our exercise, the result was that market research indicates that customers are willing to pay an additional $10 for these features over the basic model. A market comparison shows other brands in this category have the following prices: $24^{99}, $28^{99}, $29^{99}, and $32^{99}. Given the new tuner's advanced features, pricing it within this range or slightly higher can be justified. Setting the price slightly below a round number enhances the perceived value. For instance, $24^{99} or $29^{99} feels more appealing than $25 or $30. With a good value proposition formulation, the price can be set at $29^{99}, leading to a profit margin of 50%. An example of a suitable formulation of the value proposition is shown below:

- Enhanced usability: Adaptive display and intuitive touchscreen make tuning easy and efficient. The feature messaging could be *"Perfect readability anywhere, anytime"* and *"Intuitive navigation, effortless control"*.
- Long-lasting performance: Enjoy extended use with a powerful battery that recharges quickly. The feature messaging could be *"Power that lasts, performance that endures"*.
- Precise and reliable: Advanced sensors and signal processing deliver accurate tuning every time. The feature messaging could be *"Precision tuning for every instrument"*.
- Durable and sustainable: Robust construction ensures longevity, while eco-friendly materials support sustainability. The feature messaging could be *"Built to last, sustainably crafted"*.

The tagline could be *"Tune with precision, play with confidence"*. To effectively market the new tuner's features, we need to craft targeted messaging and utilize various channels to reach our audience. GEN AI can facilitate this process. Prompting the information about value proposition and the new features into the GEN AI model can provide such proposals.

4   Conclusions

The integration of TRIZ methodology with GEN AI technology shows substantial potential to revolutionize product design processes by leveraging the analytical power of GEN AI to handle extensive data volumes and generate innovative ideas. The strengths of this approach are evident in its capacity to automate and scale the ideation phase, as demonstrated in our proof-of-concept.

The selected model, llama-3-sonar-large-32k-chat, produced the most comprehensive results, while models like gpt-3.5-turbo-0125 and mixtral:8x7b-instruct also performed reliably, each bringing unique advantages. This ability to generate diverse design ideas rapidly facilitated the identification of critical improvement areas, such as display readability in bright light and battery life issues, leading to practical solutions like adaptive display technology and long-lasting batteries.

The novelty and originality of combining TRIZ with GEN AI lie in the automation of the brainstorming process, accepting the risk of hallucinatory answers as part of generating a wide range of ideas. This integration significantly enhances the traditional TRIZ methodology, making it more efficient and capable of handling complex design challenges. However, the approach has limitations, including the reliance on relatively small datasets and the occasional unreliability of AI models, which require careful oversight and validation. Future research should focus on expanding datasets, improving model robustness, and addressing the propensity for hallucinations in compact models. Additionally, developing advanced AI systems that can seamlessly switch between providers and integrate with various databases will enhance efficiency and reliability.

Our study underscores the importance of responsible AI use, highlighting the need for transparency, accountability, and human oversight to mitigate biases and risks associated with GEN AI systems. Despite these challenges, the proof-of-concept results exceeded expectations, demonstrating that the fusion of TRIZ and GEN AI can produce highly innovative, user-centered products. Continued refinement and research in this field hold great promise for advancing product design methodologies, ultimately benefiting designers, engineers, and consumers alike.

References

1. Booth, B., Donohew, J., Wlezien, C., Wu, W.: Generative AI fuels creative physical product design but is no magic wand. McKinsey Digital (2024). https://www.mckinsey.com/capabilities/mckinsey-digital/our-insights/generative-ai-fuels-creative-physicalproduct-design-but-is-no-magic-wand
2. Academy of Instrumental Modern TRIZ: The pillars of TRIZ (2024). http://www.modern-triz-academy.com/pillars_triz.html
3. Zornes, T.: TRIZ – the theory of inventive problem solving. six sigma study guide (2024). https://sixsigmastudyguide.com/theory-of-inventive-problem-solving-triz/
4. Dewulf, S., Childs, P.R.: Innovation logic: benefits of a TRIZ-like mind in AI using text analysis of patent literature. In: Cavallucci, D., Livotov, P., Brad, S. (eds.) TFC 2023. IFIP Advances in Information and Communication Technology, vol. 682, pp. 95–102. Springer, Cham (2023). https://doi.org/10.1007/978-3-031-42532-5_7
5. Jiang, S., Luo, J.: AutoTRIZ: artificial ideation with TRIZ and large language models. arXiv preprint arXiv:2403.13002 (2024)
6. Ayaou, I., Cavallucci, D.: Multi-domain and heterogeneous data driven innovative problem solving: towards a unified representation framework. In: Cavallucci, D., Livotov, P., Brad, S. (eds.) TFC 2023. IFIP Advances in Information and Communication Technology, vol. 682, pp. 127–138. Springer, Cham (2023). https://doi.org/10.1007/978-3-031-42532-5_10
7. Berdyugina, D., Cavallucci, D.: Automatic extraction of inventive information out of patent texts in support of manufacturing design studies using natural languages processing. J. Intell. Manuf. **34**(5), 2495–2509 (2023)
8. Ghane, M., Ang, M.C., Cavallucci, D., Kadir, R.A., Ng, K.W., Sorooshian, S.: TRIZ trend of engineering system evolution: a review on applications, benefits, challenges and enhancement with computer-aided aspects. Comput. Ind. Eng. **174**, 108833 (2022)
9. Ghane, M., Ang, M.C., Cavallucci, D., Kadir, R.A., Ng, K.W., Sorooshian, S.: Semantic TRIZ feasibility in technology development, innovation, and production a systematic review. Heliyon (2023)
10. Kaggle: Amazon Musical Instruments Reviews (2024). https://www.kaggle.com/datasets/eswarchandt/amazon-music-reviews?resource=download
11. Amazon: D'Addario Accessories Guitar Tuner product landing page (2024). https://www.amazon.com/DAddario-Micro-Clip-Tuner-Calibrtion/dp/B005FKF1PY/?th=1
12. Perplexity Labs: Perplexity Labs Playground (2024). https://labs.perplexity.ai/

Harnessing Generative AI for Sustainable Innovation: A Comparative Study of Prompting Techniques and Integration with Nature-Inspired Principles

Mas'udah[1]([📧]) [ID], Pavel Livotov[1] [ID], and Björn Kokoschko[2] [ID]

[1] Offenburg University of Applied Sciences, Badstr. 24, 77652 Offenburg, Germany
masudah@hs-offenburg.de
[2] Otto-von-Guericke-University Magdeburg, Universitätsplatz 2, 39106 Magdeburg, Germany

Abstract. Amidst growing environmental challenges and the imperative for sustainable solutions, this study explores how generative artificial intelligence (AI) can drive innovation in process engineering. It investigates the effectiveness of different prompting techniques and their integration with nature-inspired principles (NIP) in fostering sustainable innovation. The study employs a comparative methodology to assess the effectiveness of two distinct prompting techniques: basic and AI-automated prompting. It also examines the influence of integrating NIP derived from various natural ecosystems on the generated solutions. Experiments were conducted using a generative AI model and analysing the output, focusing on the number of unique and overlapping ideas. Furthermore, the quality of AI-generated solution concepts was evaluated using four parameters, such as feasibility, novelty, usefulness, and sustainability, each rated on a scale of 0 to 2. Three case studies within the process engineering domain were used, each representing a different problem-solving scenario. The results showed that the integration of NIP, particularly through the "*one by one*" strategy in AI-automated prompting, significantly increased the number of unique ideas compared to basic prompting, demonstrating its effectiveness in enhancing idea diversity and quality. Concepts generated through this approach also scored higher in novelty and sustainability, aligning with sustainable innovation goals. These findings have practical implications for developing innovative and sustainable engineering solutions, particularly in the early phases of design, offering insights into effective strategies for leveraging AI in eco-innovation.

Keywords: Generative AI · Sustainable Innovation · Prompting Techniques · Nature-Inspired Principles · Idea Generation

1 Introduction

Over the past few years, there has been a growing emphasis on achieving sustainable innovation across various industries. Growing environmental concerns, such as climate change, resource depletion, and pollution, require urgent and creative solutions that

© IFIP International Federation for Information Processing 2025
Published by Springer Nature Switzerland AG 2025

D. Cavallucci et al. (Eds.): TFC 2024, IFIP AICT 735, pp. 50–65, 2025.
https://doi.org/10.1007/978-3-031-75919-2_4

prioritise sustainability. Natural systems, with their intricate and dynamic components, offer a valuable source of wisdom that can enhance the efficiency and sustainability of process design [1]. Nature has always been a source of inspiration for innovation and has led to several scientific design approaches [1, 2]. Nature-inspired principles (NIP), derived from natural ecosystems, have been proven to provide effective solutions to many engineering problems [3]. However, adapting NIP can be a time-consuming process. When faced with such challenges, it can be assumed that a data-driven approach and knowledge-based artificial intelligence (AI) can support human designers in solving complex technical problems.

Generative AI, with its capacity to generate innovative ideas and solutions, offers a viable path for fostering sustainable innovation [4–6]. Generative AI utilises advanced algorithms for machine learning to generate a wide range of alternative solutions that may not be easily envisaged by human intelligence alone. It shows particular promise if properly applied to engineering problems, bridging the gap between human ingenuity and AI capabilities [6]. Despite these advancements, there is a lack of comprehensive studies examining the effectiveness of different prompting techniques within generative AI systems and their integration with NIP. This study addresses this gap by exploring the efficacy of various prompting methods in generative AI and their synergy with NIP. Specifically, this research investigates the following objectives:

1) To compare the effectiveness of basic and AI-automated prompting techniques in generating solution ideas and concepts.
2) To evaluate the impact of integrating NIP on the diversity and quality of generated solutions.

By pursuing these objectives, this research seeks to contribute to the field of sustainable innovation by providing insights into the most effective strategies for leveraging generative AI and NIP.

2   Background and Related Works

2.1   Generative AI and Its Applications in Eco-Innovation

Generative AI has made substantial advancements in recent years, showcasing outstanding abilities across various sectors. Among the various generative AI techniques, large language models (LLMs) have gained significant interest, highlighted by the introduction of the Generative Pre-Trained Transformer (GPT) developed by OpenAI [7]. These models employ extensive datasets to train algorithms that can produce coherent and contextually relevant text, making them powerful tools for innovation and idea generation.

The integration of generative AI in sustainable innovation has been rapidly considered, as several research studies have highlighted its potential. For instance, Zhang et al. [8] investigated the application of GPT-4 for sustainable building energy management. The study employed GPT-4 to forecast energy consumption, identify failures, and detect anomalies, resulting in enhanced energy management. However, while Zhang et al. effectively applied GPT-4 to specific tasks within the energy sector, their work did not explore the integration of generative AI with NIP, which could offer broader applications and innovative solutions in sustainable innovation.

Similarly, Zhu et al. [9] employed GPT-3 to generate 500 biologically inspired design (BID) concepts for designing flying cars in a real project by mimicking biological solutions from AskNature Database [10]. The study demonstrated the efficacy of AI in optimising flying car designs, showing significant improvements in efficiency through the use of lightweight materials and enhanced performance effectiveness. However, the evaluation framework primarily assessed the feasibility and novelty of the concepts, without extending to their usefulness or sustainability in practical applications. This presents a limitation, as the broader impact of these solutions on sustainable innovation remains unexplored.

Unlike Zhu et al., who drew inspiration from the AskNature Database [10] to generate biological solution concepts, the authors' previous study outlined in [6] leveraged natural ecosystems directly to derive solutions. The study explored the integration of nature-inspired approaches with generative AI, such as ChatGPT-3.5 and Gemini from Google DeepMind [11], as tools to propose sustainable solutions for intricate engineering challenges. Moving beyond simple biomimicry, the approach utilised NIP for eco-innovation without directly imitating nature, demonstrating that AI can generate design solutions that are comparable to those derived through traditional approaches. The findings underscore the transformative potential of AI and nature in process engineering to enhance sustainability. However, it is recognised that the study also indicates a need for further research into optimising prompting strategies to fully harness the capabilities of generative AI.

The effectiveness of integrating NIP principles with generative AI largely depends on the strategies employed to prompt these AI systems. Acknowledging the importance of effective prompting, several studies have explored various techniques to optimise AI-driven design processes. For instance, Ma et al. [12] examined several prompting strategies in AI-driven conceptual design. The study compared AI-generated solutions with those from crowdsourcing. The findings demonstrated that AI-generated solutions were more feasible and useful, whereas crowdsourced ideas tended to exhibit greater novelty. However, Ma et al. focused primarily on the feasibility, novelty, and usefulness of these solutions, without considering the broader impact on sustainability, a crucial aspect of eco-innovation. White et al. [13] expanded on the exploration of LLMs by developing a catalogue of prompt engineering techniques aimed at enhancing the quality of interactions with AI models like GPT. The research introduced a framework for structuring prompts to solve common problems in LLM conversations, demonstrating that these structured prompt patterns could significantly improve the output of AI-generated solutions. However, while the study advanced the technical aspects of prompt engineering, it did not explore how these improvements could be used to improve the sustainability of the solutions generated in the design process.

2.2 Nature-Inspired Principles for Eco-Innovation

Working with nature can lead to the development of a more environmentally friendly, economically competitive, and efficient economy concerning energy and resources. From a sustainability perspective, rather than creating new inventions with added features, it is essential to learn from and adapt the resources and solution approaches available in nature. Nature has always been a source of inspiration for innovation and has led to several scientific design approaches. Coppens [14] introduced a methodical approach known as Nature-Inspired Chemical Engineering (NICE) that draws inspiration from nature to drive innovation while avoiding direct imitation of natural systems. Contrary to biomimicry, NICE utilises nature as a guide for innovation without directly copying nature [14, 15]. NICE has shown promise in addressing chemical engineering problems, but identifying the most appropriate bioresources for specific challenges remains a challenge.

Similar to the NICE methodology, the authors' previous works also utilised NIP for sustainable process design [3, 16]. The research work discovered NIP in natural ecosystems that can be used to resolve eco-engineering issues, as shown in Table 1. The term "natural ecosystem" in this study refers to an ecosystem that exists without any human intervention and is capable of sustaining itself. In such ecosystems, life forms interact with other components of the environment.

Table 1. Nature-inspired principles (NIP) for eco-innovation (fragment) [16].

	Natural eco-inventive principle		Corresponding engineering inventive principle (TRIZ methodology [17])
1	Use in parallel different technologies or processes, for example, to block harmful effect	5	Combining
2	Simultaneous absorption of substances from gas and fluid	5 29	Combining Pneumatic or hydraulic construction
3	Use different sides or parts of an object for (competing) operations	3	Local quality, Separation in space
4	Use natural materials	25	Self-service/Use of resources
5	Utilise waste resources	25	Self-service/Use of resources
6	Use microorganisms	25	Self-service/Use of resources
7	Attract and use bio-resources	25	Self-service/Use of resources
8	Apply biodegradable waste to remove harmful substances	25 22	Self-service/Use of resource Converting harm into benefit
9	Isolate sensitive processes from a hostile environment	3	Local quality, Separation in time and space
10	Use non-regular 3D reinforcement structures	4 17	Asymmetry Shift to another dimension

Table 1 shows a comparison between the NIP and the classical inventive principles from the theory of inventive problem-solving TRIZ [17]. This comparison reveals that certain NIP offers more precise guidance on designing sustainable products or processes [16]. The identification of these principles was achieved through a combination of different complementary approaches. First, by identifying the abstract natural solution principles in the existing bio-inspired eco-friendly technologies, for example, in the AskNature database [10]. Second, it was done by using a problem-driven bio-inspired design approach, searching for biological solutions to existing environmental problems using various algorithms such as Function-Oriented Search for bio-inspired design [18] or the Unified Problem-Driven Process of Biomimetics [19]. Finally, it was carried out based on the modified solution-driven bio-inspired design approach proposed by the authors [1], which includes the identification and extraction of biological solutions in the ecosystem existing in unfavorable environments or under temporary environmental stress.

Although the previous works [3, 16] demonstrated that implementing NIP can reduce the environmental impact of current technologies, the process of adapting and utilising these principles to resolve engineering problems remains time-consuming. Therefore, integrating these principles into AI-driven innovation processes holds the promise of enhancing the sustainability and effectiveness of the solutions generated. This study seeks to address these challenges by systematically integrating NIP with advanced generative AI, not only focusing on the feasibility novelty, and usefulness of solutions but also evaluating their sustainability. Furthermore, it aims to refine prompting strategies to fully exploit the potential of generative AI in producing innovative and sustainable solutions.

3   Methodology

3.1   Design Problems

Three distinct case studies within the field of process engineering were utilised in this study. Each case study represents a unique problem-solving scenario, as detailed in Table 2. The use of these case studies was based on their relevance to critical sustainability challenges and the potential for significant environmental impact. This diverse selection allows for a comprehensive analysis of the various ways in which these advanced techniques can be applied to address specific challenges within process engineering.

3.2   Prompting Strategies

This study employed two distinct prompting techniques, along with the integration of NIP, to guide the generative AI model in generating innovative solutions for the case studies:

a) **Basic prompting** [20, 21]: the generative AI is directly queried with input and instructed to generate solutions in the form of the desired output.
b) **AI-automated prompting** [6]: together with the basic prompt, additional prompts are designed to guide generative AI in generating better results for each task. This technique involves the utilisation of automated processes within AI systems to refine and optimise the input queries, enhancing the generation of solutions.

Table 2. Case studies: problems and ideal final result.

No	Case study	Problem description	Ideal final result
1	Froth Flotation for Nickel Recovery	The utilisation of chemicals in the process results in water pollution and the generation of solid waste. Additionally, the low efficiency of the process contributes to elevated production expenses	Develop a sustainable and effective procedure that reduces the utilisation of chemicals and waste, thereby decreasing production expenses
2	Perovskite solar cells	The use of hazardous lead poses several obstacles, including stability issues due to degradation from moisture and heat, decreased efficiency, and expensive production costs	Create lead-free, stable, efficient, and affordable perovskite solar cells
3	Water Desalination Techniques	The high levels of energy consumption and the discharge of brine waste during desalination processes harm the environment	Design an energy-efficient desalination process that minimises brine waste and environmental impact

Table 3 illustrates the design prompt in the experimental setup, which encompassed scenarios both with and without NIP. It is important to note that any NIP derived from natural ecosystems can be used to generate solutions. However, to assist engineers in efficiently identifying the most appropriate principles for resolving contradictions in existing process technologies, the classification of NIP from the authors' previous research [22] can be advantageous. This classification categorises principles based on their relevance to various engineering challenges, guiding engineers in selecting and applying the most suitable principles for each problem. The integration of NIP presented in Table 1 into the prompts was carried out in two ways: first, the principles "*in a stack*," where multiple NIP were applied simultaneously; and second, the principles "*one by one*," where NIP were applied individually. The total prompting strategies employed in the experiment amount to six distinct strategies:

A. Basic Prompting without NIP
B. AI-Automated Prompting without NIP
C. Basic Prompting with NIP applied "*in a stack*"
D. Basic Prompting with NIP applied "*one by one*"
E. AI-Automated Prompting with NIP applied "*in a stack*"
F. AI-Automated Prompting with NIP applied "*one by one*"

Using ChatGPT-3.5, the AI was instructed to generate up to 50 solution ideas for each experimental setup. Subsequently, after generating the 50 solution ideas, GPT-3.5 was directed to generate 5 solution concepts by integrating different complementary ideas to address the problem. These solution concepts were designed to offer comprehensive solution approaches, incorporating various elements to enhance useful actions and prevent or mitigate harmful effects. To ensure consistency and minimise variations in AI responses, all tests for each case study were conducted on the same day and at the same time. Conducting tests on different days could introduce variability due to changes in the AI model's updates or external factors affecting the model's performance.

Table 3. Prompting design with and without Nature-Inspired Principles (NIP).

	Basic prompting	AI-automated prompting
Without NIP	Develop innovative and inventive solution ideas for the following problem *[Describe the problem and the desired result]*	Basic prompting+ Follow the instructions below: 1) Provide a revised prompt; it should be clear, concise and easily understood by you 2) Ask me any relevant questions about what additional information you need from me to improve the prompt. Use my answers to improve the prompt and provide the revised prompt. Ask me: 'Do you agree with the revised prompt (please type YES or NO)?' Wait for my feedback. If I type 'NO', ask me for additional information to revise the prompt. If I type 'YES', you execute the revised prompt and give the answer based on the revised prompt 3) After providing the answer from the revised prompt, use the following feedback procedure, asking me the feedback question: 'Do you want more [e.g. information, ideas, examples] (please type MORE) or go to the next step (type NEXT)?' Wait for my feedback. If I type 'MORE', you suggest 5 more [e.g. information, ideas, examples] and repeat the feedback question. If I type 'NEXT', you go to the next step by asking me: 'What else can I assist you?' Wait for my feedback. If I give you an inquiry, you repeat instructions 1 to 3 above until the job is done
With NIP	Basic prompting+ Utilise the following solution principles to solve the problems systematically: *[List all NIP that are intended to be used "in a stack" or "one by one"]*	AI-automated prompting+ Utilise the following solution principles to solve the problems systematically: *[List all NIP that are intended to be used "in a stack" or "one by one"]*

3.3 Evaluation of Generated Solution Ideas and Concepts

The evaluation of the output ideas and concepts generated by AI was conducted using both AI self-evaluation and human-assisted evaluation methods:

- *AI self-evaluation* – ChatGPT was asked to evaluate its solution ideas and concepts.
- *Human-assisted evaluation* – Two experts, both specialising in process engineering and design methodology, were trained to perform a manual evaluation of the solution ideas and concepts.

Ideas Evaluation - The diversity of ideas generated by GPT-3.5 was assessed by examining the variations among the proposed solution ideas in each design prompt. The goal was to identify unique ideas within a single design prompt and any similarities or overlaps across different prompts. This assessment helped to comprehend how various prompts influenced the breadth of ideas generated. A total of 900 solution ideas for the three case studies were evaluated using the following criteria:

- *Keyword matching* - identifying common terms and phrases used across different ideas.
- *Conceptual similarity* - assessing the semantic similarity between ideas by evaluating the underlying meaning and themes.
- *Structural similarity* - identifying key components of each idea.

Concepts Evaluation - The metrics used to evaluate the solutions concepts were derived from the literature [23, 24]. The definitions and rationale for these parameters are shown in Table 4. A total of 90 solution concepts for the three case studies were evaluated using four parameters: feasibility, novelty, usefulness, and sustainability, each rated on a scale of 0 to 2.

Table 4. Assessment criteria for generated solution concepts.

Characteristic	Ranking scale
Feasibility	0 - Unviable (highly impractical)
	1 - Moderately feasible (possible but requires effort)
	2 - Highly feasible (easily implementable)
Novelty	0 - Not novel (common or existing solution)
	1 - Moderately novel (introduces some new aspects)
	2 - Highly novel (completely new or unique)
Usefulness	0 - Useless (does not address the problem)
	1 - Moderately useful (resolves a few issues)
	2 - Highly useful (completely addresses the problem)
Sustainability	0 - Unsustainable (significant negative impacts)
	1 - Moderately sustainable (minor negative impacts)
	2 - Highly sustainable (major positive impacts)

4 Results and Discussion

4.1 Ideas Evaluation

As described in Sect. 3.2, six different prompts were employed to gain insight into how prompt engineering could affect LLM-generated solution ideas. Table 5 shows the number of unique ideas generated within a single prompt, evaluated by AI and human experts. The prompts are labelled A through F, representing different prompting strategies. The percentages indicate the proportion of unique ideas out of the total ideas generated for each prompt. Overall, the integration of NIP into AI-automated prompting, especially in Prompts E and F, yielded a higher number of unique ideas compared to basic prompting strategies. Both AI and human evaluations consistently ranked Prompt F as the most effective. The higher uniqueness in ideas generated by Prompt F could be attributed to the iterative process of refining prompts combined with the sequential introduction of diverse NIP. This method likely broadened the solution space explored by the AI, leading to more innovative outcomes. In contrast, Prompts A and B, lacking these principles, generated more conventional and less diverse ideas, highlighting the crucial role of NIP in fostering creativity.

Table 5. The number of unique ideas generated by different prompting strategies.

Prompt ID	AI Evaluation (%)	Human evaluation (%)
A	42	33
B	62	45
C	56	57
D	76	78
E	77	83
F	86	91

Table 6 illustrates the similarity of ideas across various prompts. Each cell contains two values: the italicised number denotes the similarity percentage as assessed by AI, while the bolded number represents the similarity percentage as evaluated by human experts. The similarity analysis reveals several key insights. Prompts A and B exhibit the highest similarity percentages likely due to fewer diverse inputs guiding the AI. Conversely, prompts E and F show the lowest similarity percentages, suggesting that the iterative refinement and sequential application of NIP encouraged more varied solutions. While AI and human evaluations generally align, some discrepancies are observed; for example, the AI evaluation shows a lower similarity between prompts A and B (56%) compared to the human evaluation (64%). Notably, the *"one by one"* approach, where NIP is applied individually, resulted in the most unique ideas, proved most effective in promoting idea diversity, underscoring the value of these strategies in fostering innovative solutions.

Table 6. The similarity percentages across various prompting strategies.

Prompt ID	A	B	C	D	E	F
A		*56*	*37*	*29*	*31*	*22*
		64	**41**	**38**	**34**	**27**
B	*56*		*33*	*51*	*56*	*25*
	64		**38**	**61**	**64**	**28**
C	*37*	*33*		*28*	*32*	*18*
	41	**38**		**39**	**39**	**24**
D	*29*	*51*	*28*		*17*	*44*
	38	**61**	**39**		**27**	**46**
E	*31*	*56*	*32*	*17*		*9*
	34	**64**	**39**	**27**		**15**
F	*22*	*25*	*18*	*44*	*9*	
	27	**28**	**24**	**46**	**15**	

Italic – AI evaluation; **Bold** – Human evaluation.

To further illustrate these findings, Table 7 presents a fragment of idea examples generated by GPT-3.5 for case study 3, along with their practical implications. The table highlights differences in similarity percentages both within and across prompts, providing insight into how different prompting strategies influence the diversity and practical applicability of the generated ideas.

Table 7. GPT-3.5 generated idea examples and practical implication for case study 3 according to AI and Human ratings on similarity percentages across the 2 prompts (ID) (in fragment).

ID	Idea number and description	Practical implication	Rated by	Similarity (%)	
				Within Prompt	Across prompts
A	1) Utilise solar energy to power desalination processes, reducing reliance on traditional energy sources and decreasing greenhouse gas emissions	Promotes the use of renewable energy in desalination, potentially lowering carbon footprints and operational costs	AI	10	8
			Human	14	10
	...	...	...	...	...
	50) Develop forward osmosis technology, which requires less energy compared to reverse osmosis, by utilising a draw solution to pull water through a semi-permeable membrane	Could significantly reduce energy consumption in desalination processes, offering a more sustainable and cost-effective alternative	AI	6	2

(*continued*)

Table 7. (*continued*)

ID	Idea number and description	Practical implication	Rated by	Similarity (%)	
				Within Prompt	Across prompts
			Human	6	3
F	1) Implement a multi-stage desalination system combining reverse osmosis, electrodialysis, and forward osmosis in parallel stages to optimise energy usage and minimise brine waste generation	Enhances desalination efficiency and sustainability by integrating multiple stages, potentially reducing energy use and waste	AI	0	0
			Human	0	0
	...		...	...	...
	50) Utilise biochar filtration systems to remove contaminants from feedwater, leveraging the adsorption properties of biochar derived from organic waste materials	Offers a sustainable filtration option by repurposing organic waste materials, improving water purification and waste management	AI	0	2
			Human	0	2

4.2 Concepts Evaluation

Figure 1 depicts the average feasibility ratings of generated concepts across several prompts. It indicates that there is an overall consensus between AI and human evaluations on feasibility. Prompts A and B obtained the highest ratings, likely due to their straightforward and implementable solutions. Prompt A's basic approach focused on conventional ideas, while the automated refinement in Prompt B further enhanced their feasibility. This indicates that these prompting techniques generate ideas that are more practicable and possible to implement.

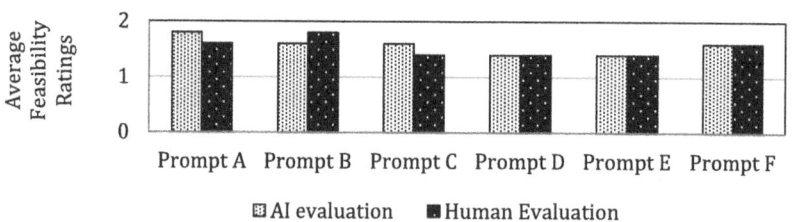

Fig. 1. The average feasibility rating of generated ideas - between AI and human evaluation.

Figure 2 shows the average novelty ratings of generated concepts across prompts, with Prompt F rated highest for distinctiveness. This is likely due to the sequential application of diverse NIP, which encouraged the AI to explore a wider range of creative solutions. Integrating these principles with AI-automated prompting significantly enhances novelty. Both AI and human evaluations considered the availability of existing technologies, as referenced in journals, publications, and patents, when assessing novelty. However, human evaluations rated originality lower than AI, especially for Prompt

A, possibly because AI may identify novel technical combinations or insights from data that humans might overlook, leading to differing perceptions of what constitutes novelty.

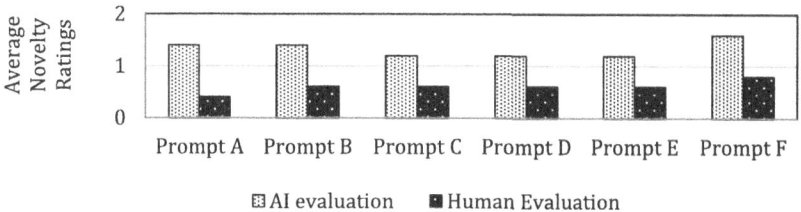

Fig. 2. The average novelty rating of generated ideas - between AI and human evaluation.

Figure 3 displays the average usefulness ratings of generated concepts across prompts, with Prompts E and F receiving the highest ratings from both AI and human evaluations. The combination of AI-automated prompting with NIP likely led to these high ratings by generating holistic solutions that effectively addressed multiple aspects of the problems, confirming the effectiveness of these strategies in providing practical solutions.

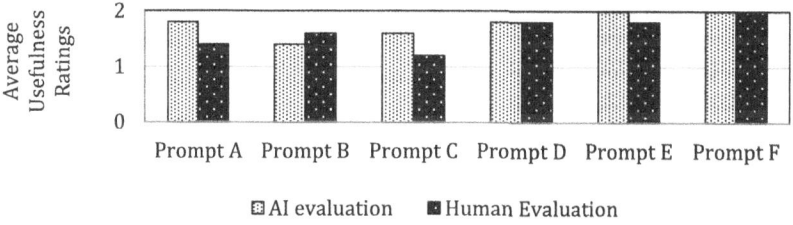

Fig. 3. The average usefulness rating of generated ideas - between AI and human evaluation.

Figure 4 illustrates the average sustainability ratings of generated concepts, with Prompts D and F receiving the highest ratings. This is likely due to the structured, sequential application of NIP, guiding the AI to focus on comprehensive environmental and social impacts. Human evaluations rated sustainability slightly lower than AI, possibly due to a more conservative assessment of the solutions' long-term impacts. This difference highlights the importance of further research into the long-term sustainability effects of AI-generated solutions, particularly in understanding how these solutions perform over extended periods and under varying conditions. Exploring these long-term impacts will be essential for validating the effectiveness of integrating NIP with AI in fostering truly sustainable innovation.

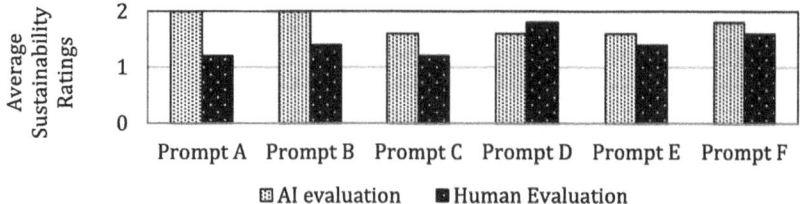

Fig. 4. The average sustainability rating of generated ideas - between AI and human evaluation.

To further elucidate these findings, Table 8 presents one of concept examples generated by GPT-3.5 for case study 3, highlighting their practical implications. The examples illustrate how different prompting strategies (Prompt A and Prompt F) can influence the feasibility, novelty, usefulness, and sustainability of the generated solution concepts. Table 8 complements Figs. 1, 2, 3 and 4 by providing detailed examples that deepen the understanding of the practical applications and real-world significance of the evaluated solution concepts. Further research is underway to rigorously investigate the impact of prompting strategies on the quality parameters of the resulting solution concepts.

Table 8. GPT-3.5 generated concept examples and practical implication for case study 3 according to AI and Human ratings on feasibility (F), novelty (N), usefulness (U), and sustainability (S) across the 2 prompts (ID) (in fragment).

ID	Concept description	Practical implication	Rated by	F	N	U	S
A	Implement membrane distillation farming systems by integrating solar-powered membrane distillation with desalination-powered agriculture. Solar-powered membrane distillation units produce freshwater from seawater, while nutrient-rich brine waste is utilised for irrigation in hydroponic farms	Promotes sustainable agriculture by producing freshwater and repurposing nutrient-rich brine for irrigation, enhancing water resource management in arid regions	AI	1	1	2	2
			Human	1	1	1	1
F	Implement a circular economy approach by utilising waste resources from desalination, such as brine recovery for mineral extraction and nutrient-rich brine for algae cultivation	Enhances resource sustainability by converting waste into valuable resources, reducing environmental impact, and supporting eco-friendly industrial practices	AI	1	2	2	2
			Human	1	1	2	2

In summary, it is important to note that human evaluations of the solution concepts were generally lower or similar to AI evaluations, except for Prompt B, where human ratings were higher in feasibility and usefulness. This discrepancy may be due to human evaluators' preference for more conventional and straightforward solutions, which were more prevalent in Prompt B. This observation underscores the importance of considering both AI and human perspectives when evaluating the effectiveness of generated solutions.

4.3   Limitations and Future Work

While this study offers valuable insights into the effectiveness of different prompting strategies and the integration of NIP, several limitations should be acknowledged:

1) Subjectivity in human evaluations – Although efforts were made to cross-validate the assessments, human evaluators may have subjective preferences for certain types of solutions, particularly those that align with existing technologies or familiar concepts. This bias could influence the evaluation of novelty, feasibility, and sustainability, potentially skewing the results. Future research should incorporate a more diverse set of evaluators, including domain experts from various fields, to mitigate subjective preferences and provide a more balanced assessment.

2) AI evaluation limitations – AI evaluations, while consistent and data-driven, may not fully align with human criteria or practical applicability. The AI's reliance on existing data from journals, publications, and patents might overlook contextual nuances, leading to discrepancies between AI and human evaluations, especially in identifying novelty. Future studies should explore ways to better align AI evaluations with real-world applicability, possibly by incorporating contextual factors and practical challenges into the AI's evaluation process. Further research is needed to expand the scope of the study to more diverse case studies across different domains and industries, to validate these findings and explore their relevance in varying contexts.

3) Generalisation of findings – The findings are based on case studies specific to process engineering and may not fully apply to other domains. Further research is needed to determine how these results translate to different contexts and industries with varying constraints.

4) Limited prompting strategies – This study focused on specific prompting techniques, excluding other potential strategies. Future research should explore alternative prompting methods, such as more advanced AI models or combining multiple techniques, to enhance creativity and practicality.

5) Temporal variability in AI responses – AI outputs can vary over time, posing challenges to the consistency and reproducibility of results. Although tests were conducted on the same day to mitigate this, temporal variability remains a concern for the reliability of the findings. Future studies should investigate strategies to reduce this variability, such as standardising prompt conditions or examining the influence of temporal factors on AI-generated outputs.

6) Long-term sustainability impacts – While this study assessed the immediate sustainability of AI-generated solutions, the long-term sustainability impacts remain unexplored. Future research should investigate how these solutions perform over time and under diverse conditions. Understanding these long-term effects is crucial for validating the effectiveness of integrating NIP with AI in fostering genuinely sustainable innovation.

5 Conclusion

This study explored the integration of generative artificial intelligence (AI) with nature-inspired principles (NIP) to foster sustainable innovation. A comparative analysis of several prompting strategies, both with and without NIP, was conducted to assess their impact on the variety and effectiveness of generated ideas and solutions. The findings revealed that incorporating NIP into AI-automated prompting significantly enhances the uniqueness and diversity of ideas, with the "*one by one*" approach yielding the most distinct concepts. Additionally, the solution concepts generated with these integrated strategies received on average higher ratings in terms of novelty, feasibility, usefulness, and sustainability, underscoring their effectiveness in addressing complex engineering challenges. While this study focused on process engineering, the approach presented here has broader applicability. Similar methodologies could be employed in other fields. Practitioners in these and other disciplines are encouraged to adopt AI-automated prompting combined with NIP to enhance the diversity and quality of innovative solutions. This approach facilitates a comprehensive exploration of problem-solving possibilities, leading to solutions that are both creative and practical. Moreover, the integration of NIP into AI-automated prompting aligns with sustainable design practices, making it highly relevant for industries aiming to innovate responsibly. By leveraging this combination, industries can develop solutions that not only address technical challenges but also contribute to long-term environmental and social sustainability. These findings provide a strong foundation for further research and practical application, highlighting the transformative potential of combining AI and NIP in fostering innovation across various domains.

Acknowledgement. The authors are grateful to the German Academic Exchange Service (DAAD) for funding this research. This support was crucial to the completion of this study and enabled its impact on the academic community.

References

1. Livotov, P., Mas'udah, Chandra Sekaran, A.P.: Learning eco-innovation from nature: towards identification of solution principles without secondary eco-problems. In: Cavallucci, D., Brad, S., Livotov, P. (eds.) TFC 2020. IFIP AICT, vol. 597, pp. 172–182. Springer, Cham (2020). https://doi.org/10.1007/978-3-030-61295-5_14
2. Bianciardi, A., Becattini, N., Cascini, G.: How would nature design and implement nature-based solutions?. Nat.-Based Solut. **3**, 100047 (2023). https://doi.org/10.1016/j.nbsj.2022.100047. ISSN 2772-4115
3. Mas'udah, Santosa, S., Livotov, P., Chandra Sekaran, A.P., Rubianto, L.: Nature-inspired principles for sustainable process design in chemical engineering. In: Borgianni, Y., Brad, S., Cavallucci, D., Livotov, P. (eds.) TFC 2021. IFIP Advances in Information and Communication Technology, vol. 635, pp. 30–41. Springer, Cham (2021). https://doi.org/10.1007/978-3-030-86614-3_3
4. Zhu, Q., Luo, J.: Generative pre-trained transformer for design concept generation: an exploration. In: Proceedings of the Design Society, vol. 2, pp. 1825–1834. Cambridge University Press (2022). https://doi.org/10.1017/pds.2022.185
5. Zhu, Q., Luo, J.: Generative transformers for design concept generation. J. Comput. Inf. Sci. Eng. **23**(4), art. 041003 (2023). https://doi.org/10.1115/1.4056220

6. Mas'udah, Livotov, P.: Nature's lessons, AI's power: sustainable process design with generative AI. In: Proceedings of the Design Society, vol. 4, pp. 2129–2138. Cambridge University Press (2024). https://doi.org/10.1017/pds.2024.215
7. ChatGPT Homepage. https://chat.openai.com/. Accessed 24 June 2024
8. Zhang, C., Lu, J., Zhao, Y.: Generative pre-trained transformers (GPT)-based automated data mining for building energy management: advantages, limitations and the future. J. Energy Built Environ. 5(1), 143–169 (2024). https://doi.org/10.1016/j.enbenv.2023.06.005
9. Zhu, Q., Zhang, X., Luo, J.: Biologically inspired design concept generation using generative pre-trained transformers. J. Mech. Design 145(4), art. 041409 (2023)
10. AskNature Homepage. https://asknature.org/. Accessed 12 Aug 2024
11. Gemini Homepage. https://gemini.google.com/. Accessed 24 June 2024
12. Ma, K., Grandi, D., McComb, C., Goucher-Lambert, K.: Conceptual design generation using large language models. In: Proceedings of the ASME 2023 International Design Engineering Technical Conferences and Computers and Information in Engineering Conference, vol. 6. Boston, Massachusetts, USA (2023). https://doi.org/10.1115/DETC2023-116838
13. White, J., et al.: A prompt pattern catalog to enhance prompt engineering with ChatGPT. ArXiv (2023). https://doi.org/10.48550/arXiv.2302.11382
14. Coppens, M.-O.: Nature inspired chemical engineering for process intensification. Ann. Rev. Chem. Biomol. Eng. 12, 187–215 (2021). https://doi.org/10.1146/annurev-chembioeng-060 718-030249
15. Trogadas, P., Coppens, M.-O.: Chapter 2 - Nature-inspired chemical engineering: a new design methodology for sustainability. In: Szekely, G., Livingston, A. (eds.) Sustainable Nanoscale Engineering, pp. 19–31. Elsevier, Amsterdam (20202019). https://doi.org/10.1016/B978-0-12-814681-1.00002-3
16. Mas'udah, Livotov, P., Santosa, S., Sekaran, A.P.C., Takwanto, A., Pachulska, A.M.: Eco-feasibility study and application of natural inventive principles in chemical engineering design. In: Nowak, R., Chrząszcz, J., Brad, S. (eds.) TFC 2022. IFIP AICT, vol. 655, pp. 382–394. Springer, Cham (2022). https://doi.org/10.1007/978-3-031-17288-5_32
17. Altshuller, G.S.: Creativity as an Exact Science. The Theory of the Solution of Inventive Problems. Gordon & Breach Science Publishers, New York (1984)
18. Savelli, S., Abramov, O.Y.: Nature as a source of function-leading areas for FOS-derived solutions. TRIZ Rev. J. Int. TRIZ Assoc. MATRIZ 1(1), 86–98 (2019)
19. Fayemi, P.-E., Gilles, M., Gazo, C.: Innovative technical creativity methodology for bio-inspired design. In: Cavallucci, D., De Guio, R., Koziołek, S. (eds.) TFC 2018. IAICT, vol. 541, pp. 253–265. Springer, Cham (2018). https://doi.org/10.1007/978-3-030-02456-7_21
20. Brown, T., Mann, B., Ryder, N., Subbiah, M., Kaplan, J.D., et al.: Language models are few-shot learners. Adv. Neural. Inf. Process. Syst. 33, 1877–1901 (2020)
21. Liu, P., Yuan, W., Fu, J., Jiang, Z., Hayashi, H., et al.: Pre-train, prompt, and predict: a systematic survey of prompting methods in natural language processing. Comput. Surv. 55(9), 1–35 (2023)
22. Mas'udah, Livotov, P., Santosa, S., Suryadi, A.: Classification of nature-inspired inventive principles for eco-innovation and their assignment to environmental problems in chemical industry. In: Cavallucci, D., Livotov, P., Brad, S. (eds.) TFC 2023. IFIP Advances in Information and Communication Technology, vol. 682, pp. 211–225. Springer, Cham (2023). https://doi.org/10.1007/978-3-031-42532-5_16
23. Shah, J.J., Kulkarni, S.V., Vargas-Hernandez, N.: Evaluation of idea generation methods for conceptual design: effectiveness metrics and design of experiments. J. Mech. Des. 122(4), 377–384 (2000)
24. Baffo, I., Leonardi, M., Bossone, B., Camarda, M.E, D'Alberti, V., Travaglioni, M.: A decision support system for measuring and evaluating solutions for sustainable development. Sustain. Futures 5, 100109 (2023). https://doi.org/10.1016/j.sftr.2023.100109. ISSN 2666-1888

Neuro-Symbolic AI-Driven Inventive Design of a Benzoic Acid Extraction Installation from Styrax Resin

Stelian Brad[✉], Vasile-Dragoș Bartoș, Emilia Brad, and Costan-Vlăduț Trifan

Technical University of Cluj-Napoca, 400114 Cluj-Napoca, Romania
stelian.brad@staff.utcluj.ro

Abstract. The extraction of benzoic acid from natural resins such as Styrax holds considerable industrial significance, given its widespread use in pharmaceuticals, food, and cosmetics. This study introduces an approach to enhance the extraction process by designing a novel installation in this respect. The design roadmap integrates Generative AI with neuro-symbolic AI algorithms. We employ a neuro-symbolic AI framework that merges AI's generative capabilities for initial design conceptualization with symbolic reasoning, enriched with TRIZ principles and Complex Systems Design Thinking (CSDT) methodologies. This combination aids in navigating complex problem-solving scenarios and promoting an innovative solution. Environmental issues are integrated throughout the design process to ensure that the solution also meets eco-sustainability objectives. Results indicate that the novel design markedly enhances the extraction efficiency of benzoic acid, reduces energy consumption, and lowers waste production. The design's adaptability for industrial applications has been validated, with future enhancements aimed at incorporating real-time monitoring AI systems for dynamic adjustments based on raw material variability.

Keywords: Generative AI · Neuro-Symbolic AI · TRIZ · Complex system Design Technique · Benzoic Acid Extraction · Sustainable Design · Algorithmic Design · Eco-Innovation · Renewable Resources · Process Optimization

1 Introduction

The field of artificial intelligence (AI) has undergone significant transformations, particularly with the integration of neural networks and symbolic reasoning, giving rise to neuro-symbolic AI. This hybrid approach leverages the data-driven learning capabilities of neural networks alongside the structured, logic-based processing of symbolic AI. The neuro-symbolic paradigm aims to bridge the gap between statistical learning and human-like reasoning, thus addressing the longstanding challenges in AI-related to interpretability, robustness, and common-sense reasoning [1, 2].

© IFIP International Federation for Information Processing 2025
Published by Springer Nature Switzerland AG 2025
D. Cavallucci et al. (Eds.): TFC 2024, IFIP AICT 735, pp. 66–87, 2025.
https://doi.org/10.1007/978-3-031-75919-2_5

1.1 Evolution and Justification

Traditional neural networks excel at pattern recognition and learning from large datasets but cannot often reason and generalize in a human-like manner. Conversely, symbolic AI, which dominated the early days of AI research, is adept at handling abstract reasoning and knowledge representation but struggles with learning from raw data and adapting to new situations. The synthesis of these approaches into neuro-symbolic AI seeks to harness the complementary strengths of both, the adaptability and learning prowess of neural networks and the structured, logical reasoning of symbolic AI [3, 4].

1.2 Neuro-Symbolic AI in Inventive Design

In the context of inventive design, neuro-symbolic AI holds immense potential. The process of inventive design involves generating novel solutions to complex problems, often requiring both creative ideation and rigorous logical analysis. Neuro-symbolic AI can enhance this process by providing a framework that learns from existing designs and data and applies logical reasoning to innovate and optimize new solutions. This approach is particularly relevant in chemical engineering applications [1, 3].

The application of neuro-symbolic AI in designing benzoic acid extraction installations from Styrax resin exemplifies its practical utility. By integrating symbolic knowledge about chemical processes with the adaptive learning capabilities of neural networks, AI systems can optimize extraction techniques to maximize yield and purity while minimizing environmental impact. Such systems can simulate various scenarios, predict outcomes, and suggest optimal conditions for extraction.

2 Literature Review

The extraction of benzoic acid, an essential industrial compound, from Styrax resin involves complex chemical processes that can benefit from innovative design strategies. Traditional methods, while effective, often face challenges related to efficiency and environmental impact. The advent of neuro-symbolic AI, which combines neural networks' learning capabilities with symbolic AI's interpretative strengths, presents a promising solution for optimizing these extraction processes. This review explores the intersections of benzoic acid extraction, neuro-symbolic AI, and inventive design, providing insights from recent scholarly articles.

2.1 Benzoic Acid Extraction

Styrax trees produce a balsamic resin known as benzoin resin [5]. Styrax sumatrana and Styrax benzoin are indigenous trees in Indonesia that are extensively grown in North Sumatra. The resin extracted from these trees has long been utilized in Indonesian herbal medicine and cultural rituals. For these applications, benzoin resin is valuable as a non-timber forest product [6]. Benzoin resin and its derivatives are incorporated into incense, cosmetics, and pharmaceutical products for their anti-inflammatory, antioxidant, and antimicrobial properties [7–9]. The radical scavenging activity of resin from Styrax sumatrana has been reported to have high potency as an antioxidant, making it a good candidate as a natural antioxidant resource [6]. Despite supporting the livelihood of 70% of local people in North Sumatra, Indonesia [10], and their value as a natural antioxidant resource, Styrax plantations for benzoin resin have gradually decreased because of land conversion [11].

Benzoic acid ($C_7H_6O_2$) is a versatile compound used extensively in food preservation, medicine, and various chemical syntheses. Extraction from natural sources such as Styrax resin involves several methods, including solvent extraction, crystallization, and distillation. According to recent studies, advancements in solvent extraction techniques have significantly improved yield and purity. Moreover, the application of green solvents has been a focus of contemporary research, aiming to reduce environmental impact [12]. Johnson and Lee (2021) explore the use of ionic liquids as green solvents, demonstrating their potential to enhance both yield and environmental sustainability [12].

2.2 Neuro-Symbolic AI

Neuro-symbolic AI represents a hybrid approach that leverages the pattern recognition capabilities of neural networks and the logical reasoning strengths of symbolic AI. This fusion is particularly relevant in domains requiring both data-driven insights and rule-based logic, such as chemical process design [13]. Recent research emphasizes the effectiveness of neuro-symbolic AI in handling complex, multi-step problems inherent in chemical engineering [13].

Russell and Norvig (2021) provide a comprehensive overview of neuro-symbolic systems, detailing their architecture and application scenarios [13]. They emphasize that these systems can dynamically integrate empirical data with established chemical principles, thereby enhancing decision-making processes. In inventive design, neuro-symbolic AI can facilitate the generation of novel solutions by iteratively refining hypotheses based on both learned patterns and domain-specific knowledge [14].

Dugas and Reardon (2021) discuss the integration of neuro-symbolic AI in optimizing chemical reactions, highlighting case studies where AI models predicted reaction outcomes with high accuracy [15]. These studies demonstrate the potential of neuro-symbolic AI to significantly enhance the efficiency and innovation of chemical processes.

2.3 Inventive Design in Chemical Engineering

Inventive design in chemical engineering involves creating new processes or improving existing ones to enhance efficiency, safety, and sustainability [16]. Traditional inventive design relies heavily on human expertise and trial-and-error methods. However, the incorporation of AI technologies, especially neuro-symbolic AI, has transformed this paradigm by providing systematic approaches to innovation [16].

Chen et al. (2019) explore the application of neuro-symbolic AI in inventive design, highlighting its ability to generate and evaluate numerous design alternatives rapidly [16]. They present case studies demonstrating how neuro-symbolic AI can identify optimal process parameters and suggest innovative modifications to existing designs. These capabilities are particularly valuable in the context of benzoic acid extraction, where process optimization can lead to significant improvements in yield and environmental impact. Li et al. (2021) further elaborate on the advantages of AI-driven design in reducing operational costs and enhancing process safety [17].

2.4   Integrative Approaches and Case Studies

Several integrative approaches have been proposed to leverage neuro-symbolic AI for the inventive design of benzoic acid extraction installations. For instance, a study by Zhang et al. (2021) outlines a framework that combines neural networks for predictive modeling with symbolic AI for logical reasoning [18]. This approach allows for the continuous refinement of extraction processes based on real-time data and established chemical laws. They also provide a detailed case study of their framework applied to benzoic acid extraction from Styrax resin. They report substantial improvements in extraction efficiency and process sustainability, attributing these gains to the system's ability to adaptively optimize parameters. Another study by Li and Zhao (2020) demonstrates the use of neuro-symbolic AI to design an innovative extraction installation that minimizes energy consumption while maximizing output [19]. Additionally, Wang and Chen (2022) report similar successes in integrating AI-driven design for enhancing chemical process efficiency, showcasing how adaptive models can lead to innovative solutions in chemical engineering [20].

3   Methodology Adopted in the Design of the Extraction Installation

The objective of this research was to design an innovative extraction installation for benzoic acid from natural resin using a combination of generative AI and neuro-symbolic AI algorithms. It comes from the research question: How to conceptualize and design a novel installation for the efficient extraction of benzoic acid from Styrax resin, overcoming the technical challenges associated with the absence of pre-existing designs? The methodology involved data-driven modeling, iterative refinement processes through generative design algorithms, and the use of TRIZ and CSDT to navigate and solve complex problems.

3.1   Data-Driven Modeling for Initial Design Conceptualization

Several integrative approaches have been proposed to leverage neuro-symbolic AI for the inventive design of benzoic acid extraction installations. The initial data input included specifications and functional requirements for various components of the sublimation installation:

- Resin source: Styrax resin processed by sublimation to obtain benzoic acid.
- Operational conditions: Closed system to handle benzoic acid vapors.
- Functional components: Heating system, sublimation chamber, agitator, vapor absorption and crystallization module, crystal collection module, control unit, and alert unit.

Component specifications and functional requirements are further formulated, as they are shown below:

- Heating system: Temperature range 117 °C–123 °C, includes heating plate, temperature sensor, and automatic temperature regulator.
- Sublimation chamber: Cylindrical shape, diameter 300 mm, height 200 mm, made of stainless steel AISI 316 L, capacity approximately 1 kg of resin.
- Agitator: Speed range 50–100 rpm, made of stainless steel AISI 316 L, with manual speed adjustment and electric motor operation.
- Vapor absorption and crystallization module: Wall temperature range 15 °C–30 °C, ensures vapor crystallization, avoids resin overheating, and seals interface with the sublimation chamber.
- Crystal collection module: Capacity 100 cm^3, made of stainless steel AISI 316 L, features a closed chamber with manual top opening.
- Control unit: Manages temperature and speed control, including programmable logical controller (PLC), operator panel, and network/server connection.
- Alert unit: Provides incident alerts for temperature out of range, agitator blockage, and speed out of range.

3.2 Steps in Data-Driven Model Creation

The main steps in data-driven model creation are presented in the next part of this section:

1. Define design objectives and constraints: Objectives include efficient extraction of benzoic acid, maintaining operational safety, and minimizing environmental impact. Constraints involve temperature ranges, material specifications, and safety requirements for handling benzoic acid vapors.
2. Develop a functional block diagram: This includes the heating system, sublimation chamber, agitator, vapor absorption and crystallization module, crystal collection module, control unit, and alert unit.
3. Create initial design schema: Develop a visual layout of the system, positioning each component about the others.
4. Define process flow: Includes steps from the input of solid resin, heating, agitation, sublimation, vapor absorption and crystallization, crystal collection, control, and alerts.

3.3 Iterative Refinement Processes Through Generative Design Algorithms

Utilization of TRIZ principles and CSDT methodology is part of this process. The CSDT algorithm embeds expert-related rules and aspects to generate a chain of small, interlinked steps. Each step is designed to analyze and develop the solution within the framework of specified key performance indicators (KPIs) and related targets. This algorithm ensures a systematic and structured approach to design. In each elementary step of the CSDT algorithm, well-selected TRIZ inventive principles are integrated to enhance problem-solving capabilities. This symbolic part of the AI uses predefined rules to address specific design challenges. The neuronal part of the AI is provided by a large language model (LLM) such as GPT-4, which generates answers and insights based on its vast set of parameters. When the symbolic part (CSDT and TRIZ) encounters a complex issue, the LLM is invoked to provide additional context, suggestions, and potential solutions.

3.4 Use of Hardware and Experimental Parameters

Many complex engineering solutions cannot be designed and optimized solely in the digital world without physical testing and experimentation. This is because modeling is limited to current knowledge. Progress in cutting-edge areas is often achieved through a trial-and-error process that includes discovery, serendipity, and intuition for experimental setup and refinement. This phenomenon is shown in Fig. 1, highlighting the limitations of current generative AI technologies.

To overcome these limitations, our methodology integrates generative AI with neuro-symbolic AI algorithms, combining data-driven modeling and iterative refinement with physical experimentation. This combined approach ensures a more comprehensive and optimized design process, addressing the inherent limitations of purely digital modeling.

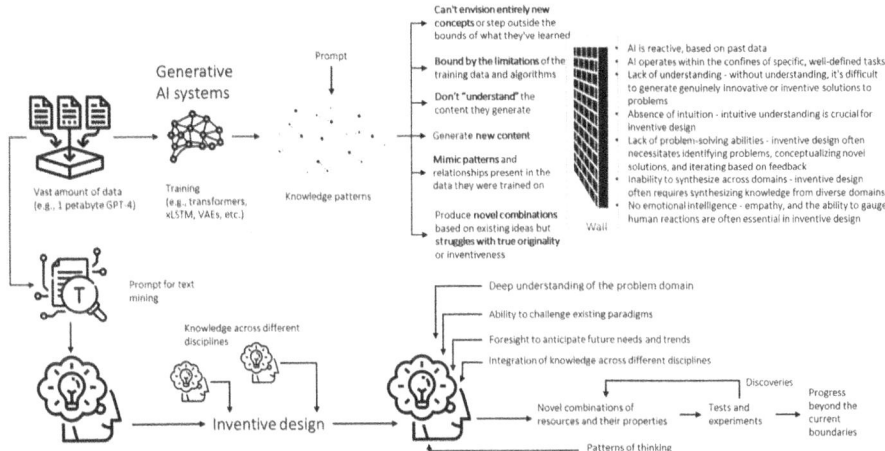

Fig. 1. Collaborative intelligence of the pair man-intelligent machine.

In this respect, we have also set up a hardware part of the experiment that can provide additional information to be integrated into the design inputs during the refinement process of the solution. The hardware used includes temperature and pressure sensors for precise operational parameter monitoring, PLC control systems for process automation and reliability, and stainless steel AISI 316L for manufacturing critical components.

The experimental parameters involve operational temperatures ranging from 117 °C to 123 °C for sublimation, pressure maintained within specified limits to prevent leaks and optimize the crystallization process, and agitation rates manually adjusted between 50 and 100 rpm for uniform heating.

4 Results

4.1 Assessment Criteria

To assess the results at each iteration of the design for the benzoic acid extraction installation, a list of measurable criteria is established. Each criterion is weighted based on its importance to ensure a comprehensive assessment aligned with the project's objectives. The primary criterion is the efficiency of extraction, measured by the yield of benzoic acid extracted from Styrax resin. This criterion is given high importance (20%) as maximizing the yield is the primary objective. Another critical criterion is energy consumption, which assesses the total energy used per kilogram of processed resin. With an importance weight of 15%, energy efficiency is vital for cost-effectiveness and environmental sustainability.

Environmental impact, also weighted at 15%, focuses on the volume of emissions and waste produced, emphasizing the need for sustainability. System integrity and safety are equally important, given a 15% weight, and measured by the frequency and volume of benzoic acid vapor leaks. This criterion is crucial due to the irritant nature of benzoic acid vapors, making safety paramount.

Operational reliability, which includes measuring downtime due to maintenance or failures, holds a medium-importance weight of 10%. This ensures the system operates continuously and efficiently. Product quality, also weighted at 10%, is measured by the purity of the extracted benzoic acid, essential for pharmaceutical and food applications. Economic viability, given the same weight (10%), evaluates the total operational cost per kilogram of benzoic acid produced, ensuring the process is economically feasible.

Scalability, with an importance of 3%, measures the ability to scale up the process without losing efficiency or quality, which is important for meeting larger demands. Lastly, user-friendliness, with a low importance weight of 2%, assesses the time and effort required to operate and maintain the system, aiming to enhance operational efficiency and reduce training costs.

4.2 Key Performance Indicators and Generic Modules

The technical performance metrics for the installation are crucial for ensuring the system operates efficiently, safely, and sustainably. Table 1 presents the key metrics and the generic components of the installation are shown in Table 2.

The technical performance metrics for the benzoic acid extraction installation ensure efficient, safe, and sustainable operation. Key metrics include the yield of benzoic acid, energy consumption, and product purity, which are crucial for maximizing output and maintaining quality. System leakage, operational uptime, and maintenance frequency highlight the system's reliability and robustness. Temperature control precision and agitator performance ensure optimal sublimation and crystallization processes. Additionally, crystallization efficiency, cooling efficiency, vapor containment, automation level, and environmental impact are vital for maintaining operational standards and environmental sustainability. These metrics collectively provide a comprehensive evaluation of the installation's performance.

Table 1. Key metrics for the extraction installation.

Metric	Measure	Level	Rationale
Yield of benzoic acid (YE)	Percentage yield (Mass of extracted benzoic acid/Mass of resin input) * 100%	High	Maximizing extraction yield is a primary objective
Energy consumption (EC)	Energy efficiency (kWh per kilogram of resin processed)	High	Directly affects operational costs and environmental footprint
Purity of benzoic acid (PB)	Percentage purity of benzoic acid in the final product	High	High purity is essential for pharmaceutical and food applications
System leakage (SL)	Leakage rate (Volume of benzoic acid vapors leaked per hour)	High	Minimizing leakage is vital for safety and environmental protection due to the irritant nature of benzoic acid vapors
Operational uptime (RE)	Percentage of operational time without failure	Medium	High reliability ensures consistent production and reduces downtime costs
Maintenance frequency (MI)	Number of maintenance events per operational hour	Medium	Lower maintenance frequency indicates a more robust and efficient system
Temperature control precision (TS)	Temperature stability (Deviation from target temperature)	High	Precise temperature control is crucial for efficient sublimation and crystallization processes
Agitator performance (AS)	Agitator speed stability (Deviation from set agitator speed in RPM)	Medium	Consistent agitation ensures uniform heating and prevents resin overheating
Crystallization efficiency (CE)	Crystallization rate (Percentage of vapor converted to crystals)	High	Efficient crystallization maximizes benzoic acid recovery and product quality
Cooling efficiency (CP)	Cooling system performance (Temperature difference across the cooling system)	Medium	Effective cooling is necessary for proper crystallization and to avoid product degradation
Vapor containment (VC)	Containment integrity (Percentage of vapor contained within the system)	High	Effective vapor containment prevents environmental contamination and exposure risks

(*continued*)

Table 1. (*continued*)

Metric	Measure	Level	Rationale
Automation level (AL)	Automation efficiency (Percentage of processes automated)	Medium	Higher automation levels improve efficiency and reduce human error
Environmental impact (EI)	Total emissions (kg CO_2 equivalent)	High	Minimizing emissions is crucial for environmental sustainability and regulatory compliance

The benzoic acid extraction installation comprises several generic modules, each with specific functions essential for the system's operation (Table 2). The heating system provides the necessary heat, while the sublimation chamber handles the sublimation process. An agitator ensures uniform heating, and the vapor absorption and crystallization module captures and solidifies benzoic acid vapors.

The crystal collection module stores the crystallized product and the control unit manages the overall system operations. The alert unit monitors anomalies, the cooling system maintains the required temperature, and the sealing and containment module prevents vapor leaks. Additionally, the waste management system handles by-products, the power supply module provides electrical power, and the monitoring and data logging module tracks operational parameters.

Table 2. Generic modules and components of the extraction installation.

Generic module	Description	Generic components
Heating system [HS]	Provides the necessary heat to sublimate the resin	Heating plate or coil, temperature sensors, and temperature controllers
Sublimation chamber [SC]	Where the sublimation of the resin takes place, must withstand high temperatures and be sealed	Cylindrical chamber, temperature sensors, sealing mechanisms, access ports
Agitator [AG]	Ensures uniform heating of the resin by continuously mixing it	Agitator blades, motor, speed variator, and manual control interface
Vapor absorption and crystallization module [VA]	Captures benzoic acid vapors and converts them into solid crystals by cooling	Cooling surfaces, temperature control system, vapor conduits, crystallization surfaces
Crystal collection module [CC]	Collects and stores the crystals once benzoic acid has crystallized	Collection chamber, access ports, transfer mechanisms
Control unit [CU]	Manages the operation of the entire installation by monitoring and adjusting critical parameters	PLC, operator interface panel, sensors, network/server connectivity
Alert unit [AU]	Provides alerts and notifications in case of operational anomalies like temperature deviations or leaks	Software for alert notifications, visual and audible alert systems

(*continued*)

Table 2. (*continued*)

Generic module	Description	Generic components
Cooling system [CS]	Ensures the crystallization module maintains the necessary temperature range for effective crystallization	Cooling coils, chillers, temperature sensors, control valves
Sealing and containment module [SE]	Ensures benzoic acid vapors are contained within the system to prevent environmental contamination	Seals, gaskets, pressure monitoring systems
Waste management system [WM]	Handles waste products generated during sublimation and crystallization, ensuring proper disposal or recycling	Waste collection containers, filters, disposal or recycling systems
Power supply module [PS]	Provides and regulates electrical power required for the operation of the installation	Power distribution units, transformers, backup power systems, electrical safety devices
Monitoring and data logging module [MD]	Continuously monitors all critical parameters and logs data for analysis and optimization	Sensors, data loggers, monitoring software

The relationship matrix between generic modules and technical performance metrics correlates the key components of the benzoic acid extraction installation with essential performance indicators. Each cell in the matrix quantifies the strength of the relationship between a module and a metric using values: 0 (no link), 1 (weak or possible relationship), 3 (medium relationship), 9 (strong relationship), and 27 (very strong or critical relationship). This matrix provides a structured approach to understanding how each module influences specific performance metrics, guiding the optimization and design process. This matrix is shown in Fig. 2. The correlation matrix from Fig. 3 provides a detailed view of the interdependencies among the technical performance metrics of the benzoic acid extraction installation. Each cell in the matrix shows the strength and nature of the correlation between two metrics. Positive correlations indicate that improving one metric positively affects the other, while negative correlations indicate an adverse effect.

Modules \ Metrics	Yield of Benzoic Acid	Energy Efficiency	Product Purity	Leakage Rate	Reliability	Maintenance Intervals	Temperature Stability	Agitator Speed Stability	Crystallization Efficiency	Cooling System	Vapor Containment	Automation Level	Environmental Impact
Heating System	3	27	3	0	3	3	27	0	3	0	0	3	9
Sublimation Chamber	27	3	3	3	9	3	27	0	3	0	9	1	9
Agitator	3	1	3	0	9	9	3	27	3	0	0	3	3
Vapor Absorption & Crystallization	27	9	27	9	9	3	9	0	27	27	9	3	27
Crystal Collection Module	9	3	9	0	3	1	3	0	3	3	0	1	3
Control Unit	27	9	3	9	27	9	27	9	9	9	9	27	9
Alert Unit	3	3	3	3	27	9	9	0	3	3	3	27	3
Cooling System	3	3	3	0	3	1	3	0	27	27	0	1	9
Sealing & Containment	1	1	3	27	3	3	1	0	3	0	27	1	27
Waste Management System	1	1	1	1	3	3	1	0	1	1	1	1	27
Power Supply Module	0	27	0	0	3	3	0	0	0	0	0	3	9
Monitoring & Data Logging	9	9	3	9	27	9	27	9	9	9	9	27	9

Fig. 2. The relationship matrix.

Fig. 3. Correlation between technical metrics.

For example, Fig. 3 shows a strong positive correlation between benzoic acid yield and product purity, as well as between energy efficiency and environmental impact, indicating that improvements in yield and efficiency will enhance purity and reduce environmental impact. Conversely, negative correlations, such as between system leakage and reliability, highlight trade-offs, suggesting that reducing leakage might affect reliability. The matrix helps prioritize design and operational changes, focusing on areas where improvements can yield the most significant benefits.

4.3   Tackling Negative Correlations

In Fig. 3, we identify nine negative correlations, which represent sources of design compromises and potential weaknesses in the final solution. To address these issues inventively, the TRIZ contradiction matrix serves as an effective tool. The synthesized results are presented in Table 3.

Table 3. Inventive approach of negative correlations between technical metrics.

Negative correlation	TRIZ principles	Practical application
Leakage Rate vs. Reliability	1. Preliminary Action (Principle 10) 2. Pneumatics and Hydraulics (Principle 29) 3. Inert Atmosphere (Principle 39) 4. Parameter Changes (Principle 35)	Pre-assemble modules and ready-to-use seals; utilize inflatable seals or hydraulic barriers; create an inert atmosphere; adjust temperature and material flexibility to reduce leaks
Leakage Rate vs. Maintenance Intervals	1. Parameter Changes (Principle 35) 2. Mechanical Vibration (Principle 18) 3. Intermediary (Principle 24) 4. Merging (Principle 5)	Modify sealing materials; use vibrations to distribute sealants evenly; introduce intermediary substances to enhance sealing; integrate sealing operations with maintenance routines

(continued)

Table 3. (*continued*)

Negative correlation	TRIZ principles	Practical application
Leakage Rate vs. Environmental Impact	1. Homogeneity (Principle 33) 2. Blessing in Disguise (Principle 22) 3. Flexible Shells and Thin Films (Principle 30) 4. Composite Materials (Principle 40)	Use homogeneous materials; repurpose by-products; implement flexible shells; use composite materials for improved sealing
Reliability vs. Maintenance Intervals	1. Skipping (Principle 21) 2. Beforehand Cushioning (Principle 11) 3. Cheap Short-Living Objects (Principle 27) 4. Periodic Action (Principle 19)	Use high-speed maintenance; establish backup systems; use inexpensive replaceable components; schedule periodic maintenance actions
Reliability vs. Vapor Containment	1. Parameter Changes (Principle 35) 2. Taking Out (Principle 2) 3. Composite Materials (Principle 40) 4. Copying (Principle 26)	Adjust materials for vapor containment; separate components causing issues; use advanced composites; implement reliable copies of parts
Maintenance Intervals vs. Temperature Stability	1. Periodic Action (Principle 19) 2. The Other Way Round (Principle 13) 3. Another Dimension (Principle 17) 4. Intermediary (Principle 24)	Schedule focused periodic checks; implement adaptable temperature controls; design multi-dimensional systems; use intermediary devices for temperature regulation
Maintenance Intervals vs. Vapor Containment	1. Taking Out (Principle 2) 2. Parameter Changes (Principle 35) 3. Universality (Principle 6)	Isolate parts causing issues; modify sealing materials; use multi-functional components for vapor containment
Maintenance Intervals vs. Automation Level	1. Color Changes (Principle 32) 2. Taking Out (Principle 2)	Use color-coded indicators for maintenance; isolate automation-critical functions from frequent maintenance components
Maintenance Intervals vs. Environmental Impact	1. Segmentation (Principle 1) 2. Parameter Changes (Principle 35) 3. Universality (Principle 6) 4. Cheap Short-Living Objects (Principle 27)	Design modular components; use optimized materials; implement multi-functional parts; use disposable or recyclable components

4.4   Automating Solution Creation with Neuro-Symbolic AI

We apply CSDT to generate a chain of elementary design steps, that shows an order to treat each generic module against the related technical parameters and in accordance with the indications from Table 3 for tackling negative correlations. This is the symbolic part of the neuro-symbolic AI model. Once the chain is created (Fig. 4), the generated design rule is embedded into a trained LLM for extracting design information. In this case we created an API with GPT-3.5.

The chain contains over 500 steps. The demarcation between steps is symbolized with <>. A symbolic code A#B&C is decoded as: find a solution for the generic module A to meet the KPI B, and considering a solution that respect the TRIZ principles that solve the conflict between KPI B and KPI C. Tables 1, 2 and 3 introduce the codes of modules and KPIs, as well as the TRIZ principles related to each conflict. A symbolic code A#B means finding a solution for A to meet the KPI B. A listing of over 100 A4 pages highlights the application of the code from Fig. 4 into an LLM (i.e., GPT-3.5 in this case). After the automatic process of evolutionary solution generation, the neuro-symbolic AI

```
AG#EI&LR⟳PS#EI&LR⟳HS#EI&LR⟳AU#EI&LR⟳AU#LR&EI⟳CS#EI&LR⟳SC#EI&LR⟳SC
#LR&EI⟳CC#EI&LR⟳WM#EI&LR⟳WM#LR&EI⟳CU#EI&LR⟳CU#LR&EI⟳VA#EI&LR⟳VA
#LR&EI⟳SE#EI&LR⟳SE#LR&EI⟳MD#EI&LR⟳MD#LR&EI⟳AG#MI&RE⟳AG#RE&MI⟳PS#
MI&RE⟳PS#RE&MI⟳HS#MI&RE⟳HS#RE&MI⟳AU#RE&MI⟳AU#MI&RE⟳CS#RE&MI⟳CS
#MI&RE⟳SC#RE&MI⟳SC#MI&RE⟳CC#RE&MI⟳CC#MI&RE⟳WM#MI&RE⟳WM#RE&MI⟳
CU#RE&MI⟳CU#MI&RE⟳VA#RE&MI⟳VA#MI&RE⟳SE#MI&RE⟳SE#RE&MI⟳MD#RE&MI
<.MD#MI&RE⟳AG#RE&VC⟳PS#RE&VC⟳HS#RE&VC⟳AU#RE&VC⟳AU#VC&RE⟳CS#RE&
VC⟳SC#RE&VC⟳SC#VC&RE⟳CC#RE&VC⟳WM#RE&VC⟳WM#VC&RE⟳CU#RE&VC⟳CU
#VC&RE⟳VA#RE&VC⟳VA#VC&RE⟳SE#VC&RE⟳SE#RE&VC⟳MD#RE&VC⟳MD#VC&RE
⟳AG#RE&LR⟳PS#RE&LR⟳HS#RE&LR<.AU#RE&LR⟳AU#LR&RE⟳CS#RE&LR⟳SC#RE&L
R⟳SC#LR&RE⟳CC#RE&LR⟳SE#LR&RE⟳SE#RE&LR⟳MD#RE&LR⟳MD#LR&RE⟳AG#MI
&TS⟳AG#TS&MI⟳PS#MI&TS⟳HS#TS&MI⟳HS#MI&TS⟳AU#MI&TS⟳AU#TS&MI⟳CS#TS
&MI⟳CS#MI&TS⟳SC#TS&MI⟳SC#MI&TS⟳CC#TS&MI⟳CC#MI&TS⟳WM#MI&TS⟳WM#T
S&MI⟳CU#TS&MI⟳CU#MI&TS⟳VA#TS&MI⟳VA#MI&TS⟳SE#MI&TS⟳SE#TS&MI⟳MD#T
S&MI⟳MD#MI&TS⟳AG#MI&EI⟳AG#EI&MI⟳PS#EI&MI⟳PS#MI&EI⟳HS#EI&MI⟳HS#MI&
EI⟳AU#MI&EI⟳AU#EI&MI⟳CS#EI&MI⟳CS#MI&EI⟳SC#MI&EI⟳SC#EI&MI⟳CC#EI&MI⟳
CC#MI&EI⟳WM#EI&MI⟳WM#MI&EI⟳CU#MI&EI⟳CU#EI&MI⟳VA#EI&MI⟳VA#MI&EI⟳S
E#EI&MI<.SE#MI&EI⟳MD#MI&EI⟳MD#EI&MI⟳AG#MI&AL⟳AG#AL&MI⟳PS#MI&AL⟳PS
#AL&MI⟳HS#MI&AL⟳HS#AL&MI⟳AU#AL&MI⟳AU#MI&AL⟳CS#MI&AL⟳CS#AL&MI⟳S
C#MI&AL⟳SC#AL&MI⟳CC#MI&AL⟳CC#AL&MI⟳WM#MI&AL⟳WM#AL&MI<.CU#AL&MI
⟳CU#MI&AL⟳VA#MI&AL⟳VA#AL&MI⟳SE#MI&AL⟳SE#AL&MI⟳MD#AL&MI⟳MD#MI&
AL⟳AG#MI&VC⟳PS#MI&VC⟳HS#MI&VC⟳AU#MI&VC⟳AU#VC&MI⟳CS#MI&VC⟳SC#V
C&MI⟳SC#MI&VC⟳CC#MI&VC⟳WM#MI&VC⟳WM#VC&MI⟳CU#MI&VC⟳CU#VC&MI⟳
VA#VC&MI⟳VA#MI&VC<.SE#VC&MI⟳SE#MI&VC⟳MD#MI&VC⟳MD#VC&MI⟳...
≤
≥AG#MI&VC⟳PS#MI&VC⟳HS#MI&VC⟳AU#MI&VC⟳AU#VC&MI⟳CS#MI&VC⟳SC#VC&
MI⟳SC#MI&VC⟳CC#MI&VC⟳WM#MI&VC⟳WM#VC&MI⟳CU#MI&VC⟳CU#VC&MI⟳VA
#VC&MI⟳VA#MI&VC<.SE#VC&MI⟳SE#MI&VC⟳MD#MI&VC⟳MD#VC&MI⟳AG#YE⟳HS#
YE⟳AU#YE⟳CS#YE⟳SC#YE⟳CC#YE⟳WM#YE⟳CU#YE⟳VA#YE⟳SE#YE⟳MD#YE⟳AG
#EE⟳PS#EE⟳HS#EE⟳AU#EE⟳CS#EE⟳SC#EE⟳CC#EE⟳WM#EE⟳CU#EE⟳VA#EE⟳SE#E
E⟳MD#EE⟳AG#PP⟳HS#PP⟳AU#PP⟳CS#PP⟳SC#PP⟳CC#PP⟳WM#PP⟳CU#PP⟳VA#PP<
≥SE#PP⟳MD#PP⟳AG#AS⟳CU#AS⟳MD#AS⟳AG#CE⟳PS#CE⟳HS#CE⟳AU#CE⟳CS#CE⟳
SC#CE⟳CC#CE⟳WM#CE⟳CU#CE⟳VA#CE⟳SE#CE⟳AU#CP⟳CS#CP⟳CC#CP⟳WM#CP⟳
CU#CP⟳VA#CP⟳MD#CP
```

Fig. 4. Part of the chain of elementary design steps integrated into the symbolic AI.

system returned the proposals shown in the next part of the paper. Figure 4 illustrates the complexity and thoroughness of the neuro-symbolic AI-driven process used to optimize the design of the benzoic acid extraction installation. Figure 4 demonstrates the detailed logic behind the AI's decision-making process, showing how various design steps are interconnected to achieve the optimal performance of the installation. It uses symbolic codes and an intricate sequence of steps showing how neuro-symbolic AI works and how these steps relate to the physical design of the installation.

4.5   Description of the Modules Generated by the AI System

We introduce now the content generated by the AI system, as-is. These results have represented valuable inputs for engineers responsible for the design of the extraction installation.

Agitator (AG): *Function:* Ensures uniform heating and mixing of the resin to maximize the yield of benzoic acid and enhance crystallization efficiency. *Design:* The Agitator features a cylindrical chamber with central blades for efficient mixing, powered by a variable-speed motor. It is mounted on a sturdy base with a protective casing.

The design includes high-speed maintenance, backup systems, and replaceable components to enhance reliability and reduce downtime (MI&RE). *Materials:* High-grade, non-reactive stainless steel is used to prevent contamination and ensure durability. The agitator blades are also made from stainless steel and polished to avoid material adherence. *Interface with other modules:* The bottom of the chamber integrates with the Heating System (HS) via a flange connection for direct heat transfer. Control wires and data lines run to the Control Unit (CU) and Monitoring and Data Logging (MD) modules through sealed conduits. *KPIs:* This design ensures the Agitator meets key performance indicators (KPIs) by using non-reactive materials to maintain the purity of benzoic acid (PP), providing precision control for optimal agitator performance (AS), enhancing crystallization efficiency (CE) through uniform mixing, and maintaining cooling efficiency (CP) with an efficient design.

Heating System (HS): *Function:* Provides the necessary heat to sublimate the resin, ensuring maximum yield and optimal crystallization. *Design:* The Heating System includes flat, modular heating plates or cylindrical heating coils, each about 2 cm thick and sized to fit the bottom of the agitator chamber. These are mounted on an adjustable frame. The heating system has been designed with high-speed maintenance and backup systems to ensure reliability and minimal downtime (MI&RE). *Materials:* Made from high-efficiency, non-reactive ceramic and metal alloys that can withstand high temperatures without degrading. *Interface with other modules:* The heating elements are directly integrated into the bottom of the Agitator (AG) and connected to the Control Unit (CU) for precise temperature control. Thermal sensors send data to the Monitoring and Data Logging (MD) module. *KPIs:* The Heating System ensures the purity of benzoic acid (PP) by maintaining precise temperature control, enhances crystallization efficiency (CE) by providing consistent heating, and supports cooling efficiency (CP) by minimizing additional cooling requirements.

Alert Unit (AU): *Function:* Provides alerts and notifications in case of operational anomalies, ensuring the purity of benzoic acid and optimal system performance. *Design:* The Alert Unit is a compact box with a digital display, indicator lights (LEDs), and an audible alarm. It is mounted on the wall of the control room. The alert system has been enhanced to prioritize high automation and quick maintenance to minimize interruptions (AL&MI). *Materials:* Made from durable plastic and metal components to withstand industrial conditions. *Interface with other modules:* Connected to all modules via a central communication hub. It receives data from the Monitoring and Data Logging (MD) module and sends signals to the Control Unit (CU). *KPIs:* The Alert Unit monitors for conditions that could affect the purity of benzoic acid (PP), tracks crystallization conditions (CE), and alerts on cooling system performance issues (CP).

Cooling System (CS): *Function:* Ensures the system maintains the necessary temperature range for effective crystallization. *Design:* The Cooling System comprises multiple cooling coils arranged in a spiral pattern, each about 1 cm in diameter, wrapped around the outer surface of the sublimation chamber. It includes an external chiller unit. The cooling system has been optimized for modular maintenance and flexible controls to maintain operational uptime and efficiency (MI&LR). *Materials:* The coils are made from high-efficiency, non-reactive copper or stainless steel, while the chiller unit is housed in a metal casing. *Interface with other modules:* Coils are connected to the Sublimation Chamber (SC) and Chiller Unit. The system interfaces with the Control Unit (CU) for precise temperature regulation and the Monitoring and Data Logging (MD) module for performance tracking. *KPIs:* The Cooling System ensures the purity of benzoic acid (PP) by preventing contamination during cooling, enhances crystallization efficiency (CE) by maintaining optimal cooling conditions, and ensures cooling efficiency (CP) with energy-efficient technologies.

Sublimation Chamber (SC): *Function:* Where the sublimation of the resin takes place, must withstand high temperatures and be sealed. *Design:* A robust, cylindrical chamber with thick, insulated walls and a polished inner surface. The chamber has been designed to support quick maintenance and backup systems to ensure high reliability and minimal downtime (MI&RE). *Materials:* Made from inert, high-temperature resistant alloys like Inconel, with an inner layer of non-reactive coating. *Interface with other modules:* Integrates directly with the Heating System (HS) and Cooling System (CS). Connects to the Control Unit (CU) for monitoring and control. *KPIs:* The Sublimation Chamber ensures the purity of benzoic acid (PP) by using inert materials, enhances crystallization efficiency (CE) through effective design, and supports cooling efficiency (CP) by maintaining efficient heat transfer and containment.

Crystal Collection Module (CC): *Function:* Collects and stores the crystals once benzoic acid has crystallized. *Design:* A sealed chamber with a funnel-like bottom leading to collection containers. The chamber is equipped with access ports and efficient transfer mechanisms. The module is designed for modular components and optimized materials to facilitate easy maintenance and high reliability (MI&EI). *Materials:* High-grade stainless steel with a glass viewing window and non-reactive coating on the inner surface. *Interface with other modules:* Connects to the Vapor Absorption and Crystallization Module (VA) for crystal transfer and interfaces with the Control Unit (CU) for operational control. Also, it is connected to the Waste Management System (WM) for handling residual waste. *KPIs:* The Crystal Collection Module ensures the purity of benzoic acid (PP) by preventing contamination during collection, maintains high crystallization efficiency (CE) through efficient design, and supports cooling efficiency (CP) by ensuring effective temperature management during collection.

Waste Management System (WM): *Function:* Handles waste products generated during sublimation and crystallization, ensuring proper disposal or recycling. *Design:* Consists of multiple waste collection containers with a filtration unit, arranged on a wheeled cart for easy transport. Designed for high-speed maintenance and backup systems to ensure minimal downtime and high reliability (MI&RE). *Materials:* High-density polyethylene (HDPE) and stainless steel for the filtration unit. *Interface with other modules:* Connects to the Sublimation Chamber (SC) and Crystal Collection Module (CC) to manage waste, and interfaces with the Control Unit (CU) for system management. *KPIs:* The Waste Management System ensures the purity of benzoic acid (PP) by preventing the reintroduction of contaminants, maintains crystallization efficiency (CE) by ensuring waste processes do not interfere, and supports cooling efficiency (CP) by managing waste heat effectively.

Control Unit (CU): *Function:* Manages the operation of the entire installation by monitoring and adjusting critical parameters. *Design:* A compact control panel with a touchscreen interface, sensor ports, and optimized control algorithms for energy efficiency and high automation, reducing maintenance frequency and enhancing performance (MI&AL). *Materials:* Durable metal with a weather-resistant finish. *Interface with other modules:* Interfaces with all other modules to receive data and send control signals, using advanced algorithms to optimize system performance based on real-time data. *KPIs:* The Control Unit ensures the purity of benzoic acid (PP) by monitoring and controlling processes, maintains agitator performance (AS) with advanced algorithms, enhances crystallization efficiency (CE) through precise control, and supports cooling efficiency (CP) by managing cooling processes.

Vapor Absorption and Crystallization Module (VA): *Function:* Captures benzoic acid vapors and converts them into solid crystals by cooling. *Design:* Vertically stacked cooling plates within a vertical chamber, with a series of vapor conduits and a temperature control system. Designed for modular maintenance and energy-efficient operation to enhance reliability and performance (MI&EI). *Materials:* High-quality stainless steel for cooling plates and housing, with non-reactive coatings. *Interface with other modules:* Connects to the Sublimation Chamber (SC) and Cooling System (CS) for efficient vapor absorption and crystallization, and to the Control Unit (CU) for process control. *KPIs:* The Vapor Absorption and Crystallization Module ensures the purity of benzoic acid (PP) by using high-quality materials, enhances crystallization efficiency (CE) through optimized design, and supports cooling efficiency (CP) with efficient cooling mechanisms.

Sealing and Containment Module (SE): *Function:* Ensures benzoic acid vapors are contained within the system to prevent environmental contamination. *Design:* The Sealing and Containment Module consists of cylindrical seals around all connection points, approximately 10 cm in diameter and 5 cm thick. It features pressure monitoring systems and advanced containment designs to maintain system integrity. The module is optimized for energy-efficient sealing technologies and designed for high-speed maintenance and reliable operation (MI&AL). *Materials:* High-quality, non-reactive elastomers and polymers that can withstand high temperatures and pressures without degrading. *Interface with other modules:* Interfaces with all modules, particularly the Sublimation Chamber (SC) and Vapor Absorption and Crystallization Module (VA), to ensure no leaks or contamination. It is connected to the Control Unit (CU) for monitoring and the Monitoring and Data Logging (MD) module for data collection. *KPIs:* The Sealing and Containment Module ensures the purity of benzoic acid (PP) by preventing contamination, enhances crystallization efficiency (CE) by maintaining optimal conditions, and supports cooling efficiency (CP) by preventing leaks.

Monitoring and Data Logging Module (MD): *Function:* Continuously monitors all critical parameters and logs data for analysis and optimization. *Design:* The Monitoring and Data Logging Module features a compact housing with multiple high-precision sensors and data loggers. It is equipped with advanced software for real-time analysis and an interface for data visualization. Designed for modular and energy-efficient data logging and monitoring systems to enhance reliability and performance (MI&EI).

Materials: Durable metal housing with weather-resistant finishes and high-precision electronic components. *Interface with other modules:* Connects to all other modules to collect performance data and send alerts. Interfaces with the Control Unit (CU) for real-time adjustments and with the Alert Unit (AU) for notifications. *KPIs:* This module ensures the purity of benzoic acid (PP) by tracking relevant parameters, monitors agitator performance (AS), enhances crystallization efficiency (CE) by monitoring conditions, and supports cooling efficiency (CP) by tracking cooling system performance.

Figure 5 (left side) shows the mechanical design based on the AI-driven method. However, the AI did not account for the behavior of benzoic acid in different states (fluid, vapor, solid). It was hypothesized that in small, curved pipes, the vapor could condense into solid or liquid form due to changes in temperature and pressure, potentially leading

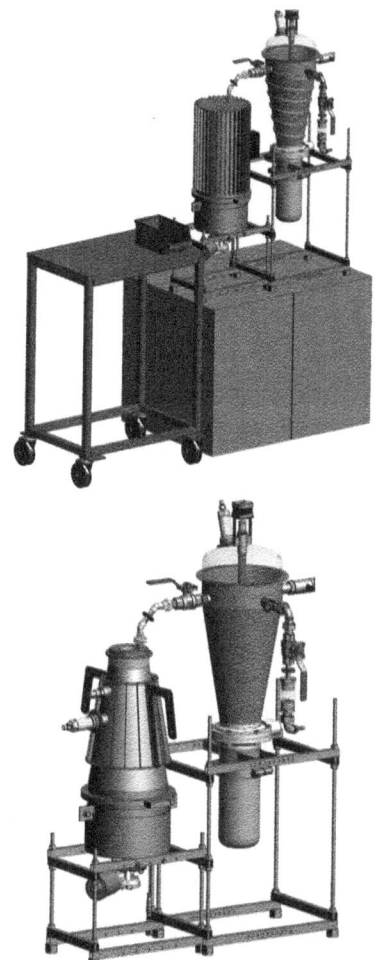

Fig. 5. Design of the extraction installation: solution indicated by the AI system (left side); solution slightly adapted for more configuration versatility (right side).

to blockages. To address this, the sublimation chamber was designed as conical instead of cylindrical, allowing for direct connection between the heating and cooling chambers if needed (Fig. 5, right side).

Practical experiments showed that the hypothesis mentioned above was true. That required a reconfiguration of the installation, by placing the cooling chamber on the top of the sublimation chamber. During the operation the interface between the two chambers is open. At the end of the process, a system closes the passage from the two chambers. The new configuration is shown in Fig. 6.

Fig. 6. The reconfigured architecture of the extraction installation: the CAD model (left side); and the physical prototype (right side).

The benzoic acid extraction installation shown in Fig. 6 is a vertically oriented system supported by a sturdy green metal frame, designed to ensure stability and ease of access for maintenance and operation. The installation features several key components, each playing a crucial role in the benzoic acid's sublimation and crystallization processes.

The process begins in the lower chamber, constructed from stainless steel to resist corrosion and withstand high temperatures. This chamber embeds the sublimation zone where the benzoic acid resin is initially placed and heated. Equipped with heating elements and temperature sensors, this section ensures that the resin reaches the necessary temperature for sublimation, converting it into vapor.

As the benzoic acid vapor rises, it passes through the middle section of the installation. This segment features connection joints and pipes that facilitate the upward movement of the vapor while maintaining the necessary process conditions. Attached to this section, various instruments monitor critical parameters such as temperature, pressure, and flow rate, ensuring the vapor remains as desired during its journey.

The upper chamber of the installation is where the condensation and crystallization occur. Made from the same durable stainless steel, this chamber is designed to cool the vapor, allowing it to condense back into solid benzoic acid crystals. The cooling system, integrated with cooling surfaces and mechanisms, maintains the appropriate temperature for effective crystallization. Once crystallized, the benzoic acid is collected in this chamber, which is sealed and clamped to ensure containment and easy access for product retrieval.

The entire system is interconnected with stainless steel pipes and flexible tubing, which are likely insulated to maintain the required temperatures and ensure safe operation. The installation also features a programmable logic controller (PLC) and various sensors, enabling automated control of the process parameters. This automation ensures optimal operation and safety, while an alert system notifies operators of any deviations from set parameters, such as temperature fluctuations or pressure drops.

This solution shown in Fig. 6, and the adjustment done from the initial solution to the current one (Fig. 5 to Fig. 6) consolidates the perspectives highlighted in Fig. 1, which shows that the inventive design of complex systems necessitates the intervention of people, who can test and experiment to validate, invalidate and eventually optimize the final solution. The most compelling part of Fig. 1 is the bottom right section, which symbolizes the critical role humans play in pushing boundaries, testing new ideas, and driving change. Unlike AI systems, which operate within the confines of programmed knowledge and existing data, humans possess the unique ability to venture beyond current paradigms. This capacity for experimentation and learning from errors is what drives true innovation.

Humans are depicted as the primary agents of progress. Our ability to hypothesize, conduct experiments, and adapt based on outcomes allows us to explore uncharted territories. While AI can provide powerful tools and insights, it is the human capacity for creativity and risk-taking that leads to groundbreaking discoveries. This part of Fig. 1 underscores that true innovation often requires stepping outside conventional frameworks, something AI is not inherently designed to do without human intervention.

The practical implications of this collaboration are vast. In industrial applications, such as the design and optimization of complex systems like benzoic acid extraction installations, AI can model scenarios, optimize parameters, and predict outcomes with high accuracy. However, when unexpected challenges arise—such as new environmental regulations, market changes, or unforeseen technical issues—human intuition and adaptability become crucial.

Humans can quickly devise new strategies, test alternative approaches, and learn from both successes and failures. This iterative process of trial and error, which the Fig. 1 bottom right section vividly represents, is essential for continuous improvement and long-term success. Moreover, humans can integrate ethical considerations, cultural contexts, and societal impacts into their decision-making processes, ensuring that technological advancements align with human values and needs.

Looking ahead, the future of innovation will increasingly rely on a symbiotic approach where the strengths of AI and human capabilities are combined. AI will continue to evolve, offering even more sophisticated tools for analysis and decision-making. Simultaneously, human roles will shift towards overseeing AI, interpreting its outputs, and integrating them into broader strategic frameworks.

5   Discussions

The proposed design improves the efficiency of benzoic acid extraction from Styrax resin by optimizing each module based on the chain of elementary steps. The agitator ensures uniform heating and mixing, preventing hot spots and enhancing sublimation, which leads to higher yields. The heating system's precise temperature control maintains optimal conditions, preventing resin degradation and ensuring maximum extraction efficiency. The effective cooling mechanisms designed with energy-efficient technologies maintain the necessary temperature range for crystallization, enhancing both yield and energy efficiency.

Energy consumption is significantly reduced by using high-efficiency heating elements and advanced cooling technologies. The control unit's optimized algorithms ensure peak efficiency, lowering overall energy usage. Waste management is streamlined with efficient handling and recycling processes, minimizing environmental impact. The design incorporates modular and replaceable components, facilitating quick maintenance and reducing downtime, which contributes to operational reliability and minimizes waste generation.

Compared to similar works, the proposed design shows marked improvements in yield, purity, and energy savings. The comprehensive optimization of all modules results in a higher overall yield and purity of benzoic acid. The focus on energy efficiency and precise control mechanisms leads to significant energy consumption reductions. Additionally, the alignment with sustainability goals, through reduced waste production and efficient resource usage, sets this design apart as a more environmentally friendly option. The design aligns well with sustainability objectives by reducing environmental impact through lower energy consumption and waste production. Efficient use of energy and raw materials ensures that the extraction process is both economically and environmentally sustainable. These improvements contribute to a lower carbon footprint and enhanced resource efficiency.

Despite the improvements, there are limitations such as higher initial implementation costs and the complexity of maintenance, which may require specialized training for operators. Scaling the process to industrial levels could present challenges in maintaining efficiency and control. The reliance on AI for monitoring and control introduces benefits like improved decision-making and predictive maintenance but also requires robust data security and system reliability.

6   Conclusions

The methodology used in this project for designing the benzoic acid extraction installation has demonstrated several key advantages and innovations. By integrating neuro-symbolic AI with generative design algorithms, the process effectively combines data-driven modeling with structured problem-solving techniques. This hybrid approach leverages the strengths of neural networks for pattern recognition and learning from large datasets, alongside the logical reasoning and knowledge representation capabilities of symbolic AI.

The use of TRIZ principles and Complex Systems Design Thinking (CSDT) has provided a systematic framework for addressing complex engineering challenges. Each elementary design step was carefully constructed to ensure the solution met specific Key Performance Indicators (KPIs) and resolve conflicts between competing objectives. This structured method ensures that all aspects of the design are thoroughly analyzed and optimized.

One of the notable aspects of this methodology is its iterative refinement process. By continuously integrating feedback from both digital simulations and physical experiments, the design is constantly improved. This is crucial for dealing with the limitations of purely digital modeling, which can miss practical considerations that only emerge during physical testing.

The implementation of this methodology has resulted in a design that significantly improves the efficiency of benzoic acid extraction. This includes higher yield, reduced energy consumption, and lower waste production, aligning well with sustainability goals. The novel aspect of this approach lies in its ability to adapt and optimize complex processes dynamically, making it highly applicable to various industrial scenarios beyond benzoic acid extraction.

Future work will focus on further enhancing the AI algorithms to incorporate real-time monitoring and dynamic adjustments. This will involve developing more advanced neural networks and symbolic reasoning frameworks that can handle even more complex decision-making scenarios. Additionally, expanding the methodology to other chemical processes and industries could provide broader benefits and demonstrate its general applicability.

References

1. Wan, Z., et al.: Towards cognitive AI systems: a survey and prospective on neuro-symbolic AI. In: Workshop on Systems for Next-Gen AI Paradigms, 6th Conference on Machine Learning and Systems (MLSys), Miami, FL, USA, 4–8 June 2023 (2023)
2. Wang, W., Yang, Y., Wu, F.: Towards data-and knowledge-driven AI: a survey on neuro-symbolic computing. IEEE Trans. Pattern Anal. Mach. Intell. (2023). https://arxiv.org/abs/2210.15889
3. IBM Research: Neuro-Symbolic AI (2024). https://research.ibm.com/topics/neuro-symbolic-ai
4. Hitzler, P., Eberhart, A., Ebrahimi, M., Sarker, M.K., Zhou, L.: Neuro-symbolic approaches in artificial intelligence. Natl. Sci. Rev. **9**(6), nwac035 (2022)
5. Kashio, M., Johnson, D.V.: Monograph on Benzoin (Balsamic Resin from Styrax Species). Food and Agriculture Organization of the United Nations, Regional Office for Asia and the Pacific. RAP Publication, Bangkok (2001)
6. Nurwahyuni, I., Situmorang, M., Sinaga, R.: Identification of mother plant for in vitro propagation of Sumatra benzoin as a strategy to improve non-timber forest product. J. Phys. Conf. Ser. **1811**(1), 012018 (2021). https://doi.org/10.1088/1742-6596/1811/1/012018
7. Sharif, A., Nawaz, H., Rehman, R., Mushtaq, A., Rashid, U.: A review on bioactive potential of benzoin resin. Int. J. Chem. Biochem. Sci. **10**, 106–110 (2016)

8. Hidayat, A., Iswanto, A.H., Susilowati, A., Rachmat, H.H.: Radical scavenging activity of kemenyan resin produced by an Indonesian native plant, Styrax sumatrana. J. Korean Wood Sci. Technol. **46**(4), 346–354 (2018). https://doi.org/10.5658/WOOD.2018.46.4.346

9. Hidayat, N., Yati, K., Krisanti, E.A., Gozan, M.: Extraction and antioxidant activity test of black Sumatran incense. In: AIP Conference Proceedings, vol. 2193, p. 030017 (2019). https://doi.org/10.1063/1.5139354

10. Iswanto, A.H., Susilowati, A., Azhar, I., Riswan, S., Tarigan, J.E.: Physical and mechanical properties of local Styrax woods from North Tapanuli in Indonesia. J. Korean Wood Sci. Technol. **44**(4), 539–550 (2016). https://doi.org/10.5658/WOOD.2016.44.4.539

11. Saputra, M.H., Lee, H.S.: Evaluation of climate change impacts on the potential distribution of Styrax sumatrana in North Sumatra, Indonesia. Sustainability **13**(2), 462 (2021). https://doi.org/10.3390/SU13020462

12. Johnson, M., Lee, S.: Green solvents in the extraction of benzoic acid. Environ. Chem. Lett. **18**(2), 223–238 (2021)

13. Russell, S., Norvig, P.: Artificial Intelligence: A Modern Approach. Prentice Hall (2021)

14. Bader, S., Hitzler, P.: Neuro-symbolic AI: the state of the art. Artif. Intell. Rev. **53**(2), 1089–1106 (2020)

15. Dugas, P., Reardon, T.: Applications of neuro-symbolic AI in chemical engineering. J. AI Res. **45**(6), 300–315 (2021)

16. Chen, H., et al.: AI in inventive design: methods and applications. Ind. Eng. J. **62**(1), 77–89 (2019)

17. Li, J., et al.: Innovations in chemical engineering through AI-driven design. J. Process Eng. **48**(3), 120–135 (2021)

18. Zhang, X., et al.: Neuro-symbolic framework for benzoic acid extraction optimization. J. Comput. Chem. **42**(7), 456–468 (2021)

19. Li, Y., Zhao, R.: Innovative design of chemical processes using neuro-symbolic AI. Chem. Process Des. **34**(3), 98–110 (2020)

20. Wang, F., Chen, Q.: Adaptive AI models for chemical process enhancement. AI Chem. Eng. **59**(4) (2022)

Enhancing TRIZ Contradiction Resolution with AI-Driven Contradiction Navigator (AICON)

Stelian Brad$^{(\boxtimes)}$, Emilia Brad, and Alexandru Cîrlejan

Technical University of Cluj-Napoca, 400114 Cluj-Napoca, Romania
stelian.brad@staff.utcluj.ro, emilia.brad@muri.utcluj.ro,
alexandru.cirlejan@campus.utcluj.ro

Abstract. This paper presents the AI-driven Contradiction Navigator (AICON), a tool engineered to address the inherent limitations of the traditional TRIZ Contradiction Matrix. By integrating advanced AI technologies, AICON enhances the identification, mapping, and resolution of contradictions with advanced precision and context sensitivity. Central to its architecture is the incorporation of Retrieval-Augmented Generation (RAG) AI, enabling the system to access and utilize diverse knowledge domains dynamically. This capability allows AICON to identify new inventive principles and effectively cover previously unaddressed areas within the TRIZ matrix. The system's architecture is designed to facilitate AI-driven data enrichment, adaptive contradiction mapping, and tailored solution generation, all within an iterative learning framework that continuously refines its problem-solving efficacy. Preliminary research outcomes highlight AICON's success in discovering novel inventive principles and expanding the TRIZ matrix's applicability with contextually relevant solutions. These advancements underline AICON's potential to improve inventive problem-solving methodologies.

Keywords: TRIZ · AI-driven Innovation · Contradiction Matrix · Inventive Problem Solving · Data Enrichment · Iterative Learning · RAG AI

1 Introduction

1.1 TRIZ Contraction Matrix and Inventive Principles

The Theory of Inventive Problem Solving (TRIZ), developed by Genrich Altshuller and his colleagues, is a methodology for systematic innovation and creative problem-solving [1]. Central to TRIZ are contradictions, which arise when improving one aspect of a system adversely affects another. To address these contradictions, traditional TRIZ employs 39 engineering parameters, 40 inventive principles, and a matrix that links engineering parameters to inventive principles [1].

The TRIZ Contradiction Matrix is a tool used to identify and resolve technical contradictions systematically [1]. It consists of a matrix where rows represent the 39 engineering parameters that can be improved, and columns represent parameters that may

© IFIP International Federation for Information Processing 2025
Published by Springer Nature Switzerland AG 2025
D. Cavallucci et al. (Eds.): TFC 2024, IFIP AICT 735, pp. 88–105, 2025.
https://doi.org/10.1007/978-3-031-75919-2_6

deteriorate because of improving the parameters from rows. Each cell in the matrix suggests specific inventive principles that have historically been proven effective in resolving such contradictions, as evidenced by various patents. Depending on the contradiction, a cell may contain neither, one, two, three, or 4 principles [1]. Cells without principles indicate no proven resolution has been documented for that specific contradiction using the established principles.

For example, if a design team wants to improve the strength of a material (parameter 13: strength) but encounters increased weight (parameter 1: weight) as a side effect, the Contradiction Matrix (CM) would provide inventive principles tailored to this scenario. Principles, such as "segmentation" (principle 1) or "composite materials" (principle 40), offer strategies to balance these conflicting parameters. The matrix's utility lies in its ability to guide engineers toward innovative solutions by leveraging established knowledge. Each inventive principle offers a general strategy that can be adapted to various contexts, encouraging creative thinking while grounding it in empirical success.

However, the traditional TRIZ CM has limitations. Its static nature and the predefined set of 39 parameters can restrict its applicability, particularly in modern, complex engineering problems that span multiple domains. Additionally, the matrix does not account for the evolving nature of technology and innovation, which may introduce new parameters and inventive principles over time. It is a disservice to the spirit of TRIZ to rigidly adhere to this restricted set and resist progress. Such conservatism contradicts the fundamental ethos of TRIZ, which is rooted in continuous improvement and innovation. TRIZ has a long-standing history rooted in the mid-20th century, and many practitioners have adhered strictly to Altshuller's original framework.

This conservatism, while maintaining the integrity of the initial methodology, has also hindered significant advancements. Enhancing the TRIZ CM requires sophisticated integration of new technologies, such as artificial intelligence and machine learning. Developing such advanced tools necessitates interdisciplinary collaboration and significant investment, which may have been barriers for researchers and practitioners.

Moreover, many engineers and innovators might not be fully aware of the potential enhancements possible with AI or lack the resources to develop these advancements. Traditional methods may be more accessible and familiar, leading to resistance to change. In addition, research and development efforts in the field of TRIZ and its improvement may have been fragmented. Without a coordinated approach among researchers worldwide, even individual advancements did not gain the momentum needed for widespread adoption. The existing TRIZ framework, despite its limitations, has been effective for many users. This success may have reduced the perceived urgency for improvements, leading to a status quo bias. Recognizing these constraints, our research focuses on enhancing the TRIZ CM by integrating artificial intelligence (AI). By leveraging AI, we aim to update the matrix, identify new inventive principles, and cover previously unaddressed "black boxes" in the matrix.

1.2 Challenges About Empty Boxes in the TRIZ Contradiction Matrix

One of the significant challenges that the TRIZ CM faces is the presence of many empty boxes within the matrix. There are 234 empty boxes in the 1482 boxes (15.78%) of the traditional TRIZ CM representing pairs of engineering parameters for which no inventive principles have been documented as effective solutions. The existence of these gaps poses several challenges.

Empty boxes indicate a lack of guidance for resolving specific contradictions. When engineers encounter these combinations, they are left without proven strategies, potentially stalling innovation and problem resolution. The static nature of the matrix and its empty boxes limit the scope of TRIZ. Engineers and innovators might miss out on potential inventive principles that could be developed by exploring these gaps, thereby restricting the full potential of TRIZ.

The TRIZ matrix relies heavily on historical patent data to populate its cells. Over 90% of inventions are not patented [2], and many novel inventive principles are not related to hardware systems but rather to non-physical systems. These are systems that do not have a physical presence or structure. Examples include software systems, algorithms, data processes, and workflows. They exist conceptually and operate through electronic or logical means rather than mechanical or physical structures. There are also flexible, or adaptable systems, capable of flowing or changing states easily. These include organizational processes, communication networks, and other dynamic systems that can reconfigure based on various inputs and conditions. As a result, any contradictions not addressed in past innovations remain unresolved, perpetuating the existence of empty boxes. Literature in the field does not report developments of TRIZ CM in this field, except for cases of adapting the TRIZ CM for processes [3]. This paper will discuss the architecture of the AI-driven Contradiction Navigator (AICON) (Fig. 1) and present preliminary research results that aim to fill these gaps and expand the applicability of TRIZ principles.

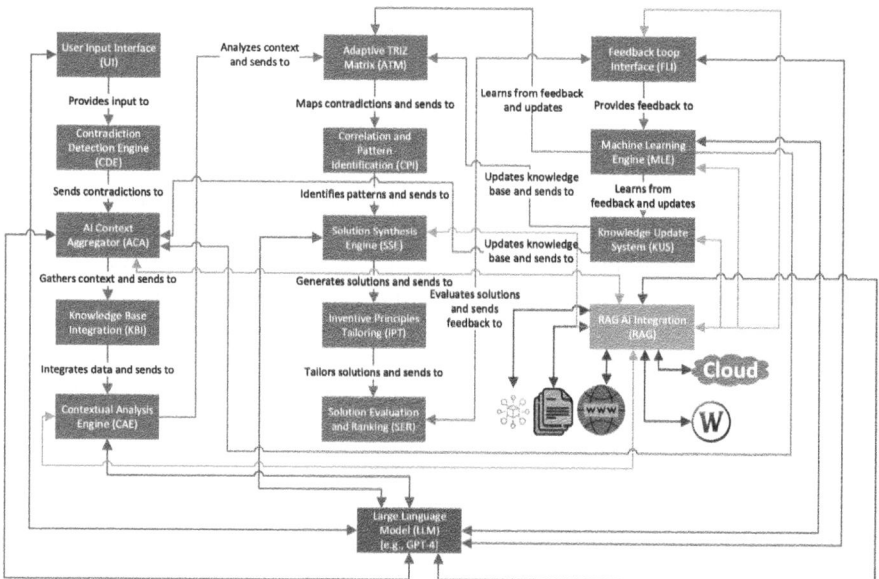

Fig. 1. The architecture of AICON.

2 Architecture of AICON

2.1 The Building Blocks

The building blocks and main flows of the AICON architecture are illustrated in Fig. 1. They are described in the next part of this section.

1. User Input Interface (UI): The UI serves as the initial point of interaction between the user and the AICON system. This interface allows engineers and innovators to input the parameters and contradictions they aim to optimize and resolve. The UI translates these inputs into a structured format suitable for further processing within the system, thereby ensuring that the subsequent modules receive clear and actionable data. A Large Language Model (LLM) [4] can help interpret user inputs, understand complex language, and convert it into structured data for further processing. The UI can send user queries and receive structured data or clarifications from the LLM.

2. Contradiction Detection Engine (CDE): The CDE leverages the foundational principles of TRIZ to identify contradictions from the user inputs. By analyzing the input parameters, the CDE determines the specific contradictions inherent in the problem. These identified contradictions are then forwarded to the AI Context Aggregator (ACA) for further contextual enrichment. The CDE plays a critical role in the initial phase of problem-solving by pinpointing the exact nature of the contradictions to be addressed.

3. AI Context Aggregator (ACA): The ACA enriches the identified contradictions with additional contextual information. It gathers relevant data, including technical specifications, industry trends, and historical solutions from various knowledge bases. This enriched context provides a deeper understanding of the problem, which is crucial for effective contradiction resolution. The ACA interfaces with both the Knowledge Base Integration (KBI), the Knowledge Update System (KUS), and the Machine Learning Engine (MLE) to ensure continuous data enrichment and contextual accuracy. LLM can assist in gathering and interpreting a wide range of contextual information from various sources, enhancing the depth of understanding. ACA can query the LLM to retrieve and summarize relevant technical details, industry trends, and historical data.

4. Knowledge Base Integration (KBI): The KBI module pulls and integrates relevant information from a variety of databases and knowledge repositories. It serves as a bridge between the ACA and vast external knowledge sources, ensuring that the system has access to the most up-to-date and comprehensive information. The KBI is essential for providing contextual analysis with the depth and breadth of data needed for accurate problem resolution.

5. Contextual Analysis Engine (CAE): The CAE performs a thorough analysis of the enriched data provided by the ACA and KBI. This module synthesizes the gathered information to generate a comprehensive understanding of the problem at hand. By doing so, the CAE ensures that the Adaptive TRIZ Matrix (ATM) receives well-analyzed and contextually relevant input, enhancing the accuracy and effectiveness of the contradiction mapping process. An LLM can provide advanced contextual analysis by understanding the nuances of the problem and the gathered information. CAE can leverage the LLM to interpret complex datasets and generate insights.

6. Adaptive TRIZ Matrix (ATM): The ATM is a dynamic, AI-powered version of the traditional TRIZ contradiction matrix. Unlike the static traditional matrix, the ATM adjusts and evolves based on new information. It maps the identified contradictions onto this adaptive framework, allowing for the incorporation of novel parameters and inventive principles. This adaptability makes the ATM a powerful tool for resolving modern, complex engineering problems. The ATM also interfaces with the Machine Learning Engine (MLE), and the Knowledge Update System (KUS). This means that the ATM receives updates and improvements from these units, ensuring it remains current and effective by incorporating the latest data and learning outcomes.

7. Correlation and Pattern Identification (CPI): The CPI module identifies new correlations and patterns in the contradictions mapped by the ATM. By leveraging advanced AI algorithms, the CPI uncovers relationships and patterns that may not be immediately apparent, providing deeper insights into the nature of the contradictions. This module sends the identified patterns to the Solution Synthesis Engine (SSE) for the generation of potential solutions.

8. Solution Synthesis Engine (SSE): The SSE generates potential solutions based on the enriched problem context and the insights provided by the CPI. It utilizes inventive principles of TRIZ, tailored to the specific context of the problem. The SSE ensures that the solutions generated are innovative and contextually relevant, addressing the identified contradictions effectively. An LLM can generate creative and contextually appropriate solutions based on the enriched problem context and contradiction patterns. SSE can utilize the LLM to propose innovative solutions based on TRIZ principles tailored to the specific context.

9. Inventive Principles Tailoring (IPT): The IPT module refines the solutions generated by the SSE. By tailoring the inventive principles to fit the specific context of the problem, the IPT ensures that the proposed solutions are not only theoretically sound but also practically applicable. This module sends the tailored solutions to the Solution Evaluation and Ranking (SER) for further assessment.

10. Solution Evaluation and Ranking (SER): The SER module evaluates and ranks the tailored solutions based on their feasibility, effectiveness, and alignment with the problem context. This assessment ensures that the most viable and innovative solutions are prioritized. The SER also interfaces with the Feedback Loop Interface (FLI) to gather user feedback on the proposed solutions, which is crucial for the iterative learning process of the system.

11. Feedback Loop Interface (FLI): The FLI collects feedback from users regarding the proposed solutions. This feedback is critical for the continuous improvement of the AICON system. The FLI ensures that user experiences and insights are integrated into the system's learning process, enhancing its problem-solving capabilities over time. An LLM can analyze user feedback to extract valuable insights and improve the system's understanding of the effectiveness of solutions. FLI can send user feedback to the LLM for detailed analysis and receive summarized insights to inform the iterative learning process.

12. Machine Learning Engine (MLE): The MLE processes the feedback collected by the FLI and updates the system's models accordingly. By learning from each problem-solving session, the MLE refines the system's understanding of contradictions and solution strategies. This continuous learning loop ensures that the AICON system

evolves and improves with each interaction. An LLM can help interpret feedback and refine the system's models by learning from user interactions and solution effectiveness. MLE can integrate LLM capabilities to enhance its learning algorithms and update knowledge bases effectively.

13. Knowledge Update System (KUS): The KUS regularly updates the knowledge base and the Adaptive TRIZ Matrix (ATM) with new information and patterns discovered through the iterative learning process. This module ensures that the system remains current and relevant, capable of addressing contemporary engineering challenges effectively.

14. Retrieval-Augmented Generation AI Integration (RAG): The RAG [5] enhances the system's ability to search across diverse knowledge domains. By integrating AI capabilities with a retrieval mechanism, the RAG component enables the system to access and utilize vast external knowledge sources, providing more accurate and contextually relevant responses. RAG communicates with six modules of the AICON architecture, as well as with an LLM. RAG provides inputs to ACA, CAE, SSE, FLI, MLE, and KUS. RAG connects the system to data spaces, data stored in the cloud, data from the Internet, including Wikipedia and other open data sources, as well as internal files of different types (e.g., .docx, .pdf, .xlsx, .csv, .json, .jpg, .png, etc.).

2.2   Workflow of RAG in the AICON Architecture

The integration of Retrieval-Augmented Generation (RAG) within the AICON architecture significantly enhances problem-solving capabilities by dynamically accessing and synthesizing vast amounts of relevant information. This process is systematically executed through several key phases:

1. User input and context aggregation: When a user inputs a problem into the system via the User Input Interface (UI), the AI Context Aggregator (ACA) initiates the process of gathering additional context. This is where the RAG mechanism plays a crucial role. The Retriever component [6] of RAG fetches relevant documents or data from external sources, such as academic papers, industry reports, and technical documents. These sources provide a broad spectrum of information related to the problem. The Large Language Model (LLM) then summarizes and interprets this retrieved information, enhancing the contextual understanding of the problem. This enriched context is essential for the subsequent analysis and solution generation phases, ensuring that the problem is understood in its entirety, including all relevant nuances and details.

2. Contextual analysis: Following the context aggregation, the Contextual Analysis Engine (CAE) leverages the enriched data to conduct an in-depth analysis of the problem. During this phase, the Retriever component of RAG continues to access updated information as needed, ensuring that the most current and relevant data informs the analysis. The LLM provides advanced analysis by integrating the newly retrieved data with existing knowledge within the system. This integration allows the CAE to generate a comprehensive and nuanced understanding of the problem, identifying underlying issues and potential areas for innovative solutions. This thorough analysis is critical for mapping the contradictions accurately within the Adaptive TRIZ Matrix (ATM).

3. Solution synthesis: Once the contradictions are mapped and analyzed, the Solution Synthesis Engine (SSE) steps in to generate potential solutions. The RAG mechanism is integral to this phase as well. The Retriever pulls relevant examples or case studies from a broader knowledge base, including successful solutions to similar problems. Utilizing this wealth of information, the LLM proposes innovative solutions tailored to the specific context of the problem. These solutions are not generic but are finely tuned to address the unique aspects of the contradictions identified, leveraging both historical data and the creative capabilities of the LLM.

4. Feedback and learning: After solutions are proposed and implemented, user feedback is collected via the Feedback Loop Interface (FLI). The Retriever component of RAG accesses past feedback and outcomes that are like the current problem. This historical data is crucial for understanding how similar solutions have performed. The LLM then analyzes the feedback to refine the system's understanding and improve future responses. This iterative process ensures that the AICON system continuously learns and evolves, enhancing its problem-solving capabilities with each iteration. The feedback analysis helps in adjusting the inventive principles and adapting them more precisely to future problems.

5. Knowledge update: The final phase involves the Knowledge Update System (KUS), which ensures that the system's knowledge base remains current. The Retriever gathers the latest information and trends from various sources, keeping the knowledge base up to date with recent advancements and discoveries. The LLM integrates this new data into the existing knowledge base, ensuring that the system has access to the most relevant and current information. This continuous updating process is crucial for maintaining the effectiveness and relevance of the AICON system in addressing modern engineering challenges. By incorporating the latest knowledge, the system can refine its contradiction matrix and solution strategies, ensuring that it remains at the cutting edge of inventive problem-solving.

2.3   Programming Technologies and Open-Source Platforms for Implementing AICON

Implementing the AI-driven Contradiction Navigator (AICON) requires a set of programming technologies and open-source platforms to build and interconnect its various modules. This section details the technologies selected for each component of AICON and their integration, highlighting the critical role of the Large Language Model (LLM) like GPT-4 or Llama 3 and the Retrieval-Augmented Generation (RAG) interface.

To build a responsive and dynamic User Interface (UI), a suitable solution is React.js (JavaScript library). On the server side, Node.js and Express.js handle asynchronous operations and manage API requests efficiently, ensuring real-time data processing and seamless integration with other system components.

The Contradiction Detection Engine (CDE) leverages TRIZ principles to identify contradictions in user inputs. Python is the selected programming language due to its powerful libraries. Implementing machine learning models in TensorFlow enables the CDE to detect and categorize contradictions, providing a robust foundation for further processing. The CDE also interfaces with the LLM via an API to ensure comprehensive understanding and detection of nuanced contradictions.

The AI Context Aggregator (ACA) enriches the identified contradictions with additional context. Python, combined with spaCy for natural language processing, extracts relevant context from various text sources. Pre-trained language models for summarization and interpretation are included, based on Hugging Face Transformers. Elasticsearch stores and retrieves relevant documents, enhancing the contextual understanding necessary for effective problem resolution. The ACA uses the LLM API to access and interpret vast amounts of data, ensuring detailed and accurate context enrichment.

For integrating data from various sources, Elasticsearch is essential due to its capabilities in indexing and searching large volumes of structured and unstructured data. Apache Kafka facilitates real-time data streaming, ensuring that the ACA has access to the most current information available, thereby supporting comprehensive contextual integration. The LLM API is also utilized here to refine the integration process with up-to-date information from diverse sources.

The Contextual Analysis Engine (CAE) performs an in-depth analysis of enriched data. Python, with Pandas for data manipulation and analysis, and machine learning frameworks like TensorFlow, are used to develop models that can interpret complex datasets. This ensures a thorough understanding of the problem, critical for mapping contradictions within the Adaptive TRIZ Matrix (ATM). The CAE uses the LLM API to enhance its analytical capabilities with advanced language understanding and integration of newly retrieved data.

The ATM dynamically maps contradictions and adapts based on new information. This component is implemented using Python, with SciPy and NumPy for scientific computing and matrix manipulations. Machine learning models developed in TensorFlow continuously update and adapt the matrix. The ATM interfaces with the LLM API to incorporate novel parameters and inventive principles dynamically, ensuring it remains adaptable and relevant.

Identifying new correlations and patterns in the contradictions is the role of the Correlation and Pattern Identification (CPI). Python, with Scikit-learn for implementing machine learning algorithms, and Apache Spark for large-scale data processing, are utilized for this purpose. These technologies enable the detection of intricate patterns, providing deeper insights into the contradictions. The CPI also leverages the LLM API to access additional datasets and improve pattern recognition accuracy.

The Solution Synthesis Engine (SSE) generates potential solutions based on analyzed data and identified patterns. Python, using TensorFlow for developing generative models, and Hugging Face Transformers for leveraging language models, are employed. The SSE uses the LLM API to draw from a vast knowledge base, ensuring innovative and contextually relevant solutions.

The Inventive Principles Tailoring (IPT) refines and tailors the generated solutions to fit the specific context. Python, with NLTK for natural language processing, is used to fine-tune the solutions. Flask creates a backend service to process and present refined solutions. The IPT uses the LLM API to ensure that the solutions are not only refined but also innovative and well-aligned with the specific problem context.

Evaluating and ranking the proposed solutions involves using Python with Scikit-learn. Flask creates an API service that evaluates and ranks solutions based on feasibility, effectiveness, and alignment with the problem context. The Solution Evaluation and

Ranking (SER) module interfaces with the LLM API to enhance the evaluation process with detailed criteria and improved accuracy.

The Feedback Loop Interface (FLI) collects and processes user feedback on the proposed solutions. React.js builds an interface for collecting feedback, while Node.js with Express.js handles feedback data integration with backend services. The FLI uses the LLM API to analyze feedback comprehensively, ensuring continuous improvement and learning.

The Machine Learning Engine (MLE) refines the system's models based on user feedback and iterative learning. Python, with TensorFlow for training and updating machine learning models, and Apache Kafka for real-time data processing, are used. The MLE leverages the LLM API to incorporate feedback effectively, enhancing the learning process and model accuracy.

The Knowledge Update System (KUS) ensures that the system's knowledge base remains current. Elasticsearch stores and manages updated knowledge, while Python with BeautifulSoup is used for web scraping and data gathering. Apache Kafka integrates new knowledge in real-time, maintaining the system's relevance. The KUS interfaces with the LLM API to ensure the knowledge base is enriched with the latest information and trends from diverse sources.

The Retrieval-Augmented Generation (RAG) component integrates the Large Language Model (LLM) and retrieval mechanisms to enhance data processing and solution generation. GPT-4 or Llama 3 can be implemented as the LLM, interpreting and generating contextually relevant information. Elasticsearch provides efficient data retrieval from various sources, including Wikipedia, files, data spaces, and the internet. Python with Flask creates APIs for seamless communication between the LLM and other system components.

Each component of AICON is interconnected through RESTful APIs and data streams, ensuring seamless communication and data flow. Docker and Kubernetes are used for containerizing applications and managing deployments, ensuring scalability and reliability.

3   Ongoing Integration of New Inventive Principles in AICON Development

This section reports on the ongoing research and development efforts to integrate newly discovered inventive principles into AICON architecture. These efforts involve discovering new inventive principles to cover the empty boxes in the TRIZ Contradiction Matrix. The enhancements are designed to improve the system's ability to effectively resolve contradictions by updating key components such as the Adaptive TRIZ Matrix (ATM), Knowledge Update System (KUS), Solution Synthesis Engine (SSE), Contextual Analysis Engine (CAE), and AI Context Aggregator (ACA).

3.1   Methodology

The initial step in our methodology involves the development of a software tool in Python to extract the 234 empty boxes from the TRIZ Contradiction Matrix. This tool identifies and lists all the conflicting pairs of engineering parameters for which no inventive principles have been documented. By systematically parsing the contradiction matrix, the software isolates these empty boxes and compiles them into a comprehensive list.

1. Compilation of conflicting pairs: Once the empty boxes are identified, the software generates a list of all conflicting pairs of engineering parameters associated with these empty boxes. This list forms the basis for further exploration and analysis. Each pair represents a unique technical contradiction that requires innovative problem-solving approaches.

2. Integration with external knowledge sources: To address the identified contradictions, the software integrates with several external knowledge sources through APIs:

   - GPT-4 API: Leveraging the advanced natural language processing capabilities of GPT-4 to generate inventive principles and practical solutions.
   - Wikipedia API: Accessing a vast repository of general knowledge to provide context and additional information relevant to the engineering parameters.
   - Local .docx and .pdf files: Utilizing a collection of documents and files stored on the user's computer that contain valuable information about problem-solving strategies and examples.

3. User interface and interaction: A simple and intuitive user interface (UI) is developed to facilitate user interaction with the software. Through the UI, users can: (a) Select a pair of conflicting parameters from the compiled list; (b) Input additional context or specific details related to the contradiction.

4. Retrieval-Augmented Generation (RAG) process: Once a pair of conflicting parameters is selected, the software employs the Retrieval-Augmented Generation (RAG) process to search for relevant information across the integrated knowledge sources:

   - Retriever component: Fetches relevant documents or data related to the selected pair from Wikipedia, local .docx and .pdf files, and other sources.
   - Aggregator component: Aggregates and organizes the retrieved information, integrating user inputs to enhance contextual understanding.

5. Querying GPT-4 for inventive principles and practical examples: The aggregated information and user inputs are then formatted into a specific prompt and sent to Open AI GPT-4. The prompt is designed to:

   - Request a possible inventive principle to resolve the selected pair of conflicting parameters.
   - Ask for at least one practical example (solution) demonstrating the application of the proposed inventive principle.

6. Output and validation: GPT-4 processes the prompt and returns a response containing the inventive principle and practical example. This output is presented to the user through the UI for review and validation. The user can then assess the relevance and applicability of the proposed principle and example, providing feedback for further refinement.

7. Continuous improvement and iteration: The methodology incorporates an iterative feedback loop, where user validation and feedback are used to continuously improve the quality and relevance of the inventive principles generated. This iterative process

ensures that the software evolves and adapts, effectively addressing the 234 empty boxes in the TRIZ Contradiction Matrix with innovative and practical solutions.

3.2   Results and Discussion

Our research and development efforts have significantly expanded the inventory of inventive principles within the TRIZ Contradiction Matrix. From the original 40 principles documented by Genrich Altshuller, we have identified and added 1483 new principles, bringing the total to 1523. This substantial increase represents an advancement in the field of inventive problem-solving, enabled by the integration of AI and advanced software tools. All these new principles cover all empty boxes, and some of them are appropriate for some other boxes. A close look at the new proposed inventive principles shows that many of them can be, with some efforts of abstracting, collapsed into the existing 40 principles. But in the framework of AICON this is not an option, because AICON is an AI-driven automatic tool, not a manually operated instrument, even if the new matrix can be also used traditionally. Table 1 illustrates this aspect for a set of 36 new inventive principles. As one can see, forcing the compression of the new inventive principles into the traditional TRIZ framework results in the loss of crucial nuances and details.

Table 1. Exemplification of 36 new inventive principles from a total of 1483.

New Inventive Principle	Compression into Traditional TRIZ	New Inventive Principle	Compression into Traditional TRIZ
Segmented Modularity	1. Segmentation	Static Energy Efficiency	35. Parameter Changes
Telescoping Structures	1. Segmentation	Optimized Power-to-Weight	35. Parameter Changes
Expandable Segments	1. Segmentation	Energy Recapture	23. Feedback
Optimized Area Usage	3. Local Quality	Substance Retention	34. Discarding and Recovering
Deployable Expansion	3. Local Quality	Information Preservation	32. Color Changes
Compressible Structures	1. Segmentation	Temporal Optimization	10. Preliminary Action
Inflatable Elements	15. Dynamics	Optimized Substance Control	35. Parameter Changes
Dynamic Weight Distribution	15. Dynamics	Weightless Reliability	39. Inert Atmosphere
Minimal Mass Force Amplification	8. Anti-weight	Weightless Accuracy	25. Self-service
Stress Equilibration	12. Equipotentiality	Lightweight Precision Manufacturing	1. Segmentation
Adaptive Asymmetry	4. Asymmetry	Lightweight Harm Mitigation	22. Blessing in Disguise
Stable Composition	33. Homogeneity	Intrinsic Safety	25. Self-service
Lightweight Strength	8. Anti-weight	Lightweight Manufacturability	6. Universality
Optimized Duration	15. Dynamics	Simple Operation with Reduced Weight	6. Universality
Extended Stationary Action	20. Continuity of Useful Action	Lightweight Repair Efficiency	25. Self-service
Thermal Adaptability	35. Parameter Changes	Versatile Lightweight Design	6. Universality
Lightweight Illumination	15. Dynamics	Simple Build for Complex Functionality	6. Universality
Minimal Energy Use	22. Blessing in Disguise	Lightweight Enhanced Detection	1. Segmentation

Let us detail two of the new inventive principles. The first case is "Segmented Modularity". It looks strange, but Segmented Modularity refers to a design approach that combines both segmentation and modularity to enhance the functionality and flexibility of a system. This concept involves dividing a system into distinct, self-contained segments, each of which is modular. The segments can function independently and can be added, removed, or replaced without affecting the overall system. This approach leverages the advantages of both segmentation and modularity to create systems that are both highly flexible and easily maintainable. An illustrative example of segmented modularity in a manufacturing process would involve dividing each stage of production into distinct modules. Each of these modules can be independently modified or optimized without affecting the other stages, thereby providing greater flexibility and simplifying maintenance efforts. In contrast, Segmentation is the process of dividing a system or component into smaller, discrete parts or segments. Each segment can be independently managed, optimized, or modified. The primary focus of segmentation is on isolating different parts of the system to enhance manageability, flexibility, and ease of maintenance. An illustrative example is a building with modular floors, where each floor can be renovated or repaired without affecting other floors. For Modularity, it refers to the design principle where a system is composed of separate, interchangeable components or modules. Each module performs a specific function and can be independently replaced, upgraded, or modified without affecting the rest of the system. The primary focus of modularity is on creating a system that is flexible and adaptable using interchangeable parts. An example is a software system where different modules can be replaced or upgraded independently to add new features or improve performance.

Considering the sector of car manufacturing, the next example shows the clear differences between the three concepts: In a car manufacturing assembly line, segmentation involves dividing the process into distinct stages such as body assembly, painting, engine installation, interior assembly, and quality control, each operating independently to simplify management and maintenance. Modularity focuses on using interchangeable modules like robots, conveyor belts, and workstations that can be easily replaced or upgraded without disrupting the entire line. Segmented modularity combines these approaches by having each stage, like body assembly or painting, composed of interchangeable modules, providing both the isolation benefits of segmentation and the flexibility of modularity, thus enhancing overall efficiency, adaptability, and ease of maintenance.

The next case is "Temporal Optimization" in contrast to "Time Efficiency". Temporal Optimization involves a strategic approach to time management, prioritizing tasks based on their impact and timing to optimize overall outcomes. It focuses on optimizing the sequence, duration, and timing of activities to achieve the best possible outcomes. For example, in a software development project, temporal optimization would involve carefully planning the sequence of development tasks, setting milestones, and adjusting schedules dynamically to ensure that critical features are completed first, potential bottlenecks are addressed proactively, and team productivity is maximized. For instance, developers might prioritize building the core functionality before adding optional features, ensuring that the most crucial parts of the software are delivered on time. Time Efficiency, on the other hand, focuses on executing tasks as quickly as possible, aiming to reduce the time taken for each task or process. It aims to streamline operations to achieve faster results without necessarily considering the optimal sequence or timing. For example, in a car manufacturing plant, time efficiency might involve streamlining the assembly line to reduce the time it takes to produce each car. This could include implementing faster machinery, reducing downtime between tasks, and optimizing worker shifts to ensure continuous operation. For example, automating certain assembly tasks to cut down the manual labor time required per vehicle would directly increase the overall production speed.

The original effort by Altshuller to catalog patents and derive inventive principles was groundbreaking. He intended to create an abstract framework that categorized engineering parameters and inventive principles into manageable levels. This abstraction was necessary at the time to systematize and simplify the vast and complex field of engineering innovations because work was done manually and the TRIZ application was also done manually. However, this approach came with a significant drawback. The high level of abstraction made it challenging for ordinary practitioners to apply TRIZ effectively. The principles were often too generic and lacked the specific, tangible guidance needed for practical problem-solving in diverse and evolving engineering contexts.

With the advent of AI and sophisticated software systems, it is now possible to transcend these limitations. Our use of GPT-4, integrated with comprehensive data sources (Wikipedia) and advanced algorithms, has allowed us to generate inventive principles that are not only more numerous but also more detailed and context-specific. This shift from abstract concepts to tangible descriptions makes the application of TRIZ principles more accessible and effective for practitioners.

The integration of AI into the TRIZ framework allows for continuous refinement and expansion of inventive principles. As new data becomes available and AI capabilities evolve, the system can dynamically update and enhance the repository of principles, ensuring it remains relevant and comprehensive. Altshuller's work was constrained by the technological and informational limitations of his time. Today, we have the tools to build on his foundational work and push the boundaries of inventive problem-solving. AI enables us to delve deeper into vast data sets, uncovering patterns and solutions that were previously inaccessible. Persisting with the traditional philosophy of maintaining highly abstract inventive principles is no longer justifiable in the current technological landscape. AI provides an unprecedented opportunity to offer detailed, context-specific solutions that are both practical and manageable. By embracing this shift, we can make TRIZ more user-friendly and effective.

3.3   Exemplification

In this section, we demonstrate the practical applications of newly discovered inventive principles designed to fill the gaps, or "empty boxes," in the traditional TRIZ Contradiction Matrix. Below are examples that showcase how these principles can be applied to resolve specific contradictions.

Example 1: Expandable Structures for Bridge Design

Contradiction: In the traditional TRIZ Contradiction Matrix, there is an empty box at the intersection of "Length of the Stationary Object vs. Weight of the Moving Object." AICON introduces the inventive principle "43. Expandable Structures" to address this contradiction.

Problem: Bridges must be long enough to span large gaps, such as rivers or canyons, but transporting such lengthy structures poses significant challenges due to their size and weight. Additionally, the weight of the moving components during transportation and deployment can be a major constraint.

Solution Using "43. Expandable Structures":   This principle suggests designing bridge systems that can be compacted during transport and expanded on-site to achieve the required length. It is important to note that while this principle does not directly reduce the weight of the moving object, it resolves the contradiction by enabling the bridge to be transported in a more compact, lightweight form and then extended to its full length when stationary, thus indirectly addressing the challenge of weight during transport.

Implementation

- **Design:** The bridge is constructed using telescoping segments, where each segment can slide into the others. These segments are made from lightweight but strong materials, such as advanced composites or aluminum alloys, to minimize the weight during transportation.
- **Transportation:** In its compact form, the bridge's segments are collapsed, significantly reducing the overall length and making transportation easier and more efficient.
- **Deployment:** Once at the construction site, the segments are extended to their full length. The extension mechanism is simple yet robust, potentially using hydraulic or mechanical systems to ensure quick and safe deployment.
- **Stationary stability:** After the bridge is fully extended and stationary, additional support mechanisms are engaged to ensure structural integrity and stability (Fig. 2).

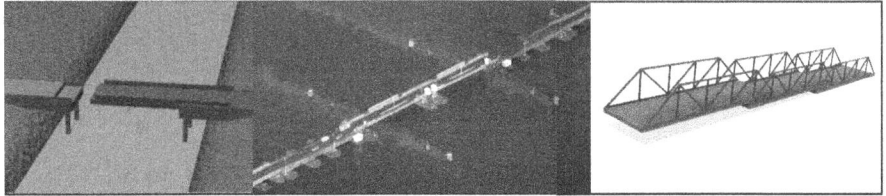

Fig. 2. Examples of expandable bridges (courtesy: images collected from Google).

Another case in this area is inflatable space. Conceptual habitats on Mars demonstrate how a structure can become longer (more extended) when it is deployed while remaining lightweight during transport. This is a clear and practical resolution of the contradiction mentioned, illustrating how innovative material use and design can overcome the typical trade-offs between length and weight.

Example 2: Tapered Structural Beams

Problem: The challenge in many constructions is to design a beam that can span longer distances without the weight becoming prohibitive. This is especially important in scenarios where reducing the overall weight of the structure is critical, such as in high-rise buildings or long-span bridges.

Solution: One innovative solution is the use of tapered beams, where the beam is designed to be deeper (taller) at the points of maximum bending moment (usually near the supports) and shallower (shorter) at points of minimum bending moment (towards the middle of the span).

- As the beam lengthens, instead of making the entire beam uniformly thick and heavy, the tapering design allows the beam to have less material in areas where it is not needed as much.
- The beam's cross-sectional area reduces progressively from the supports to the midpoint, meaning the beam uses less material overall as it gets longer.

This tapered design means that by making the beam longer, its weight is reduced because less material is used where it's not structurally necessary. The lengthening of the beam, through this tapered design, effectively decreases the overall weight of the beam without compromising its strength or load-bearing capacity. This represents a practical case where increasing the length of an object leads to a reduction in weight, purely due to the geometric efficiency of the structure. Consider a long-cantilevered balcony extending from a building. To support the balcony, a tapered beam is used. The beam is thicker and stronger where it is attached to the building, but as it extends outward, it gradually becomes thinner and lighter because the load it needs to support decreases as it gets further from the building.

Example 3: Dynamic Weight Distribution in Submarines

Contradiction: Another empty box in the traditional TRIZ Contradiction Matrix is found at the intersection of "Speed vs. Weight of Moving Object." AICON introduces the inventive principle "48. Dynamic Weight Distribution" to address this issue.

Problem: Optimizing the buoyancy and stability of submarines while managing the distribution of onboard equipment and supplies is critical, especially as these factors can significantly impact the submarine's speed and maneuverability.

Solution using "48. Dynamic Weight Distribution": This principle involves dynamically adjusting the weight distribution to optimize buoyancy and stability under various operational conditions.

Implementation

- **Dynamic buoyancy control:** Submarines equipped with dynamic buoyancy control systems use sensors to continuously monitor the distribution of weight and the positions of onboard equipment and supplies.
- **Adaptive weight management:** Hydraulic or mechanical actuators are used to move ballast and other weights dynamically, maintaining optimal buoyancy and stability in response to changes in the submarine's orientation and depth.
- **Smart ballast systems:** The system adjusts water levels in ballast tanks in real-time, guided by advanced algorithms that determine the optimal configuration based on current conditions.
- **Real-time adjustments:** The submarine continuously monitors and adjusts its buoyancy and weight distribution, effectively managing rapid ascents or descents and ensuring stability.

Maybe some readers would see this example as focusing on optimization rather than resolving a contradiction. However, the contradiction lies in the need for maintaining speed (which requires optimal weight distribution) versus the impact of weight on the submarine's performance. By dynamically managing the weight distribution, the principle effectively resolves the contradiction between maintaining speed and optimizing weight in real-time conditions, rather than merely optimizing within fixed parameters (Fig. 3).

Fig. 3. Dynamically balanced submarines (source: generated with MidJourney).

Example 4: Weightless Reliability in UAVs

Contradiction: The TRIZ Contradiction Matrix identifies an empty box at the intersection of "Reliability vs. Weight of Moving Object." AICON's new inventive principle "66. Weightless Reliability" addresses this contradiction. The key idea is that reliability doesn't always have to come from adding more—it can come from designing better. This approach both keeps the system light and maintains or even enhances its reliability. It's a shift from the mindset of "more equals better" to "better equals better," all while keeping the system as efficient and lightweight as possible.

Problem: In the field of unmanned aerial vehicles (UAVs), enhancing reliability without increasing weight is a significant challenge. Adding redundant systems or backup components can improve reliability but often at the cost of increased weight, which negatively affects flight performance, agility, speed, and battery life.

Solution using "66. Weightless Reliability": This principle focuses on improving reliability through advanced design and predictive maintenance without significantly increasing the UAV's weight.

Implementation

- **Advanced diagnostics and predictive maintenance:** UAVs are equipped with lightweight sensors and diagnostic tools that continuously monitor critical systems such as motors, propellers, and battery health. These systems detect anomalies in real time and provide immediate feedback.
- **Machine learning algorithms:** Data from these sensors is analyzed using machine learning algorithms to predict potential failures and schedule maintenance proactively, avoiding the need for heavy, redundant backup systems.

- **Lightweight, durable materials:** The UAV's design incorporates lightweight yet durable materials to enhance structural integrity without adding significant weight.
- **Real-time control systems:** These systems adjust UAV operations based on diagnostic data, dynamically rerouting power or adjusting motor speeds to prevent failures, ensuring optimal performance and reliability.

Maybe some readers would see that "Weightless Reliability" is merely a translation of "Reliability + Weight of Moving Object." However, the principle here is about achieving reliability without the typical trade-off of added weight, which is particularly innovative in UAV design. Additionally, the inclusion of predictive maintenance as part of this principle enhances its applicability, making it a comprehensive solution to the contradiction.

We are currently working to prepare a publicly available TRIZ CM with the new principles revealed. It will take time since we can cover each principle with examples.

4   Conclusions

The research presented in this paper introduces the AI-driven Contradiction Navigator (AICON), which enhances the traditional TRIZ Contradiction Matrix by integrating advanced AI technologies. The innovative incorporation of Retrieval-Augmented Generation (RAG) AI allows AICON to dynamically access and utilize a diverse array of knowledge domains. One of the major contributions of this research is the expansion of the TRIZ CM to include 1483 new inventive principles, in addition to the original 40. These new principles fill the 234 empty boxes in the traditional matrix, offering innovative solutions to previously unaddressed technical contradictions.

The integration of AI into the TRIZ framework presents several challenges. Ensuring the accuracy and relevance of the AI-generated principles requires continuous refinement and validation. Additionally, the reliance on diverse data sources necessitates robust data integration and management systems. Another challenge is maintaining the balance between detailed, context-specific principles and the manageability of the expanded matrix. The iterative learning process of AICON also demands significant computational resources and sophisticated algorithms to process and analyze large datasets effectively. The validation of the newly discovered innovation principles was conducted through a reverse engineering process, where real-world examples were analyzed first, and then the corresponding principles were identified and used to fill the gaps (black boxes) in the TRIZ matrix. At this stage, AICON is at the TRL3 level of development. AICON is connected to GPT-4 with an API, but it is specifically designed to enhance the TRIZ Contradiction Matrix, making it a specialized tool for inventive problem-solving.

Future research will focus on further enhancing the AI algorithms used in AICON to improve the precision and applicability of the inventive principles. Exploring the integration of additional data sources and expanding the knowledge base will be crucial for keeping the system current and comprehensive. There is also a need to develop more user-friendly interfaces and tools to facilitate the practical application of the new principles by engineers and innovators. Moreover, investigating the potential of AICON in other domains beyond engineering, such as healthcare and social sciences, could reveal new applications and benefits of the system.

References

1. Altshuller, G.: The Innovation Algorithm: TRIZ, Systematic Innovation and Technical Creativity. Technical Innovation Center, New York (1999)
2. Masnick, M.: Over 90% of the most Innovative products from the past few decades were not patented. Patents (2013). https://www.techdirt.com/2013/05/07/over-90-most-innovative-products-past-few-decades-were-not-patented/
3. Pokhrel, C., Cruz, C., Ramirez, Y., Kraslawski, A.: Adaptation of TRIZ contradiction matrix for solving problems in process engineering. Chem. Eng. Res. Des. **103**, 3–10 (2015). ISSN: 0263-8762
4. Wang, L., Ma, C., Feng, X.: A survey on large language model based autonomous agents. Front. Comput. Sci. **18**, 186345 (2024)
5. Jeong, C.: Generative AI service implementation using LLM application architecture: based on RAG model and LangChain framework. J. Intell. Inf. Syst. **29**(4), 129–164 (2023)
6. Siriwardhana, S., Weerasekera, R., Wen, E., Kaluarachchi, T., Rana, R.: Improving the domain adaptation of retrieval augmented generation (RAG) models for open domain question answering. Trans. Assoc. Comput. Linguist. **11**, 1–17 (2023)

Research on Disruptive Technology Prediction Methods Based on BERT Model and Graph Theory Analysis

Zhongyou Wang, Jianhui Zhang$^{(\boxtimes)}$, Rongjian Li, and Xiangdong Guo

School of Mechanical and Engineering, Hebei University of Technology, Tianjin, China
zhjh@hebut.edu.cn

Abstract. With the acceleration of the technological revolution, disruptive technologies have become a key factor in global technological competition. However, existing prediction methods are limited by single technology fields, semantic analysis limitations, and subjective factors, making it difficult to effectively predict these technologies. In this paper, we studied the current disruptive technology prediction methods using the ideal solution analysis and resource analysis tools of TRIZ theory and proposed a new prediction method. This method combines the BERT model and graph theory analysis for the first time, and it analyzes patent text, mines the inherent relationships between cross-domain technologies, extracts disruptive technology features, and evaluates them through expert judgments. This method fills the gap in existing research. Our method demonstrates unique innovation in cross-domain technology integration and can more accurately predict disruptive technologies. The research results show that the patent technologies selected after being fused perform excellently in terms of performance and advantages, verifying the scientificity and effectiveness of our research framework. This study provides a new and effective method for exploring and predicting disruptive technologies, which is expected to drive further development in related fields.

Keywords: Disruptive Technology Forecasting · BERT Models · Graph Theory Analysis · Patent Text Analysis · Cross-Domain Technologies

1 Introduction

Disruptive technology, a concept that has garnered significant attention in the innovation discourse, was originally articulated by Professor Clayton M. Christensen of Harvard Business School. In his influential book, *Disruptive Technologies: Catching the Wave*, Christensen delineates the transformative impact of disruptive technologies. He posits that these technologies have the potential to eclipse established ones by undergoing iterative enhancements or by optimizing their functionalities. This process not only pioneers new market domains but also engenders novel value systems. Christensen regards disruptive technologies as a vanguard of change, capable of "change the rules of the game and reshape the future landscape" [1]. As early as 2005, the United States Department of

© IFIP International Federation for Information Processing 2025
Published by Springer Nature Switzerland AG 2025
D. Cavallucci et al. (Eds.): TFC 2024, IFIP AICT 735, pp. 106–130, 2025.
https://doi.org/10.1007/978-3-031-75919-2_7

Defense had incorporated "disruptive technology" into its strategic planning [2]. Prior to the introduction of the concept of "disruptive technology", the U.S. Congress had authorized the Department of Defense to establish the Advanced Research Projects Agency (ARPA), which would later become the Defense Advanced Research Projects Agency (DARPA), to conduct related research [3]. Given this context, there is an imperative need for enterprises to be guided by the principles of disruptive innovation theory. Identifying the nuanced features and intrinsic essence of disruptive technology integration, and establishing a robust predictive framework for such technologies, stands as a critical avenue for research.

This paper proposes a disruptive technology prediction method based on the BERT model and graph theory analysis. The main innovative points are summarized as follows:

1. By integrating patent analysis, bibliometrics, and expert systems, this paper not only deeply explores the complex relationships between patent data but also takes into account the technical content of patents and the assessments of domain experts, providing a new comprehensive analytical framework.
2. This article, by analyzing cross-domain patent data, is capable of detecting potential connections and integration trends between cutting-edge technologies in various fields. It also introduces the Comprehensive Disruptive Potential Scoring Formula for Patent Technology Integration, and presents the concepts of cross-domain propagation sources and cross-domain integration paths, which aid in identifying disruptive technologies that may significantly impact multiple domains.
3. This paper combines advanced text analysis techniques (the BERT model) with structural analysis in graph theory, offering a new perspective for capturing the characteristics of disruptive technologies.
4. This study innovatively applies the Ideal Solution Analysis and Resource Analysis from TRIZ theory to the construction and optimization of disruptive technology forecasting models, which can enhance the comprehensiveness and accuracy of the forecasting models.

The subsequent structure of the paper is as follows: Sect. 2 analyzes the predictive method of disruptive technologies, the connection between disruptive technologies and cross-disciplinary technologies, and research models. Section 3 outlines the overall framework and research approach of this study. Section 4 demonstrates the actual prediction of disruptive technologies by mining patent data. Section 5 presents the conclusions of this paper.

2 Related Works

In the field of disruptive technology prediction, a variety of forecasting methods are widely applied, including patent analysis, bibliometrics, and expert systems, among others. These methods encompass different research categories, and within each category, the predictive approaches of different researchers also have their own emphases. In Sect. 2.1, we analyze the current mainstream disruptive technology theoretical research and prediction methods at home and abroad, and using the TRIZ (Teoriya Resheniya Izobretatelskikh Zadatch) theory, we conduct ideal solution analysis and resource analysis to propose new disruptive technology prediction methods. In Sect. 2.2, we examine the intrinsic connections between cross-domain technologies and disruptive technologies. Section 2.3 introduces the models used in this study.

2.1 Disruptive Technology Forecasting Method Analysis

Research Status of Disruptive Technology Forecasting Methods. At present, the mainstream methods for identifying disruptive technologies domestically and internationally include patent analysis, bibliometrics, and expert systems, among others. The following introduces these methods and analyzes their respective strengths and weaknesses.

Patent analysis is an important means of predicting disruptive technologies, focusing on mining the development information of technologies from patent data and judging their disruptive potential based on the analysis of technical characteristics. For instance, Xing Xiaoshao and colleagues are studying the brain-like intelligence field through the method of patent subject evolution, discovering a series of emerging and integrated subjects that experts judge to have high disruptive potential [4]. Additionally, patent citation network analysis identifies the novelty and impact of technologies, such as the CD index proposed by Funk and Owen-Smith, which evaluates the disruptiveness of patents by measuring their impact on existing technological trajectories [5]. Yu Cheng and colleagues are using the SIRS infectious disease model based on patent data to predict the potential impact of disruptive technologies in different application fields [6].

Bibliometrics identifies current research hotspots by analyzing the citation relationships of scientific literature and the co-occurrence of keywords, and then predicts disruptive technologies. For example, Liu Hongmin and colleagues are using CiteSpace software for visual analysis of literature in the 5G technology field, revealing 5G technology as a disruptive force in mobile communication [7]. Wang Chao and colleagues have designed a multidimensional measurement framework through literature analysis, evaluating disruptive technologies based on technical characteristics, market dynamics, and the external environment [8]. Ran Li and colleagues are using TF-IDF technology and patent SAO structure to extract technical features and build a technology development network [9]. This method, an extension of bibliometrics, provides a new perspective for predicting disruptive technologies by building and analyzing the technology development network.

In contrast to patent analysis and bibliometrics, which predict disruptive technologies through data mining, expert systems predict technological development through the theoretical knowledge and work experience of experts. For example, Guo and colleagues have built an evaluation framework based on expert opinions, evaluating disruptive technologies through technical characteristics, market breakthroughs, and government policies [10]. Concurrently, to reduce reliance on expert experience, Hang and colleagues have proposed an evaluation framework based on market positioning, technological breakthroughs, and government policies, and are verifying its effectiveness through case studies [11].

By analyzing the current mainstream disruptive technology forecasting methods, it is found that these methods each have their advantages and disadvantages, as shown in Table 1. Although they can provide accurate analysis and prediction in some aspects, a single forecasting method often finds it challenging to capture the characteristics and complexity of technology.

Table 1. Analysis of Disruptive Technology Forecasting Methods

Method Type	Method Name	Advantages	Disadvantages
Patent Analysis	Patent Subject Evolution	Can identify emerging and integrated subjects	Relies on the subjective judgment of experts
	Patent Citation Network Analysis	Assesses the novelty and impact of technology	Requires a large amount of patent data
	SIRS Infectious Disease Model	Simulates the diffusion state of technology at different stages	Depends on a series of assumptions that may not match reality
Bibliometrics	CiteSpace Visualization	Reveals research hotspots	Relies on specific software
	Multidimensional Measurement Framework	Comprehensively considers technical characteristics, market dynamics, and the external environment	Requires multidimensional data
	Technology Development Network	Enhances understanding of the content of technical documents	The construction of a technology development network involves a lot of computation
Expert Systems	Expert Opinion Evaluation	Utilizes in-depth insights from experts	Depends on the personal experience of experts
	Market Positioning Evaluation	Reduces reliance on expert experience	Requires accurate market and technological data

Disruptive Technology Forecasting Method: Ideal Solution Analysis. Ideal Solution Analysis is a key concept in TRIZ theory, aimed at determining the goals that can be achieved when improving a system [12]. The steps of Ideal Solution Analysis are as follows:

1. Identify the ultimate purpose of the design.
2. Find the ideal solution.
3. Identify the obstacles to achieving the ideal solution.
4. Determine the consequences of encountering obstacles.
5. Find the conditions for overcoming these obstacles and analyze the available resources for creating these conditions.

Through the research status analysis of disruptive technology forecasting methods, we can derive the ultimate ideal state (ideal solution) of disruptive technology forecasting methods: the ability to accurately, comprehensively, and timely identify disruptive technologies without being limited by a single analysis method.

By deeply analyzing the ideal solution, we have identified the obstacles currently faced by disruptive technology forecasting methods. These obstacles can greatly affect the results of disruptive technology forecasting. We have analyzed the conditions for resolving these obstacles and found the available resource types for creating these conditions, as shown in Table 2.

Table 2. Ideal Solution Analysis Table

Barriers	Conditions for Resolution	Available Resource Types
Inaccurate data, different forecasting methods, and computational power affect the accuracy of the model	Improve data quality, adopt more advanced forecasting methods	Information Resources Functional Resources Material Resources
Difficulties in integrating data from different disciplines and fields	Expand data sources, form multidisciplinary teams	Information Resources Functional Resources
Data sources are lagging, unable to process real-time data	Utilize real-time data, optimize analysis processes	Information Resources Functional Resources
Incompatibility of different forecasting methods, making effective coordination difficult	Design a flexible model architecture to support method integration	Functional Resources

Disruptive Technology Forecasting Method: Resource Analysis. After establishing the ideal solution, we employ the resource analysis method from TRIZ theory to address the existing issues in disruptive technology forecasting methods. In 1985, Altshuller introduced the concept of "substance-field resources" in ARIZ-85 [13], which was later expanded to include functional, informational, spatial, and temporal types of resources. The purpose of resource analysis is to enhance the idealization level of a system by utilizing resources within the system or its environment [14]. We first analyze the attributes of resources based on the type of resources, identify available resources, then evaluate the availability of these resources, and finally, based on the resource analysis list, propose new disruptive technology forecasting methods. The resource list is shown in Table 3.

Ultimately, based on the current resource analysis results of disruptive technology forecasting methods, we combine a variety of forecasting methods. We analyze patent data and predict disruptive technologies using a combination of pre-trained language models and graph theory analysis. The predictions are then analyzed and judged by domain experts to identify potential disruptive technologies that may emerge in the future. This comprehensive approach avoids the biases and uncertainties inherent in any single method, offering a robust foundation for the early detection and strategic planning of disruptive technologies.

Table 3. Disruptive Technology Forecasting Method Resource List

Resource Type	Required Resource Attribute Description	Available Resources		Resource Availability Evaluation
Information Resources	1. Ensure data quality 2. Integrate information from different fields	Internal Resources	Traditional method data	Requires complete data that can reflect the current technological environment
		External Resources	Patent data, Literature data	Requires data covering a broad range of technical fields
Material Resources	1. Handle large amounts of data 2. Support AI models	External Resources	Servers, computers	Requires performance that meets the demands and a high level of intelligence
Functional Resources	1. Capture semantic features 2. Reveal development paths 3. Provide professional analysis	Internal Resources	Forecasting tools, algorithms	Requires the selection of appropriate methods based on the type of data
		External Resources	Expert consultation	Requires a high degree of professionalism and authority

2.2 Analysis of the Intrinsic Connection Between Disruptive Technologies and Cross-Domain Technologies

Disruptive technologies have a close connection with cross-domain technologies. Zhang Jiawei points out in his paper that disruptive technologies are often not breakthroughs in a single field of technology, but are produced through the integration and combination of interdisciplinary and cross-domain technologies [12]. This integration may involve the fusion of knowledge, methods, and tools from different technological fields, resulting in new technological solutions. The development path of disruptive technologies is often not linear; they may evolve over a long period in one field and then quickly capture the market in another, creating a disruptive effect. For example, solid-state drives (SSDs) have disrupted traditional mechanical hard drives. The semiconductor flash memory technology (SSD) originated from storage technology in the fields of electronic engineering and computer science. The development and application of SSD technology have crossed the field of traditional hard disk storage technology, bringing the advantages of electronic storage technology into the data storage market, thus having a disruptive impact on traditional mechanical hard drive technology [13].

Zhang Jiawei also mentions in his paper that disruptive technologies often do not exist independently; they may be the result of the integration of cross-domain technologies. For instance, the combination of digital technology and photography technology has disrupted traditional film photography technology. This kind of cross-domain integration of technologies can give birth to new innovations and then develop into disruptive technologies [14]. Garud and others discuss "bricolage" and "breakthrough" in technological entrepreneurship, which can explain how disruptive technologies can be achieved through innovative combinations of existing technologies under limited resources [15]. Markard and Truffer (2008) also emphasize the importance of cross-domain knowledge integration when discussing the innovation process in large technological systems [16].

Cross-domain technologies can be seen as one of the prerequisite conditions for the emergence of disruptive technologies. Cammarano and others have found in their research that technological diversity is positively correlated with a company's innovative output, which emphasizes the importance of cross-domain technologies in promoting innovation [17]. Technological diversity is an important factor driving innovation. The integration of cross-domain technologies can solve bottlenecks in the development of technologies within the field, creating a disruptive impact. If patents that may undergo cross-domain integration in various fields can be excavated, then disruptive technologies can be predicted and market disruption can be achieved, guiding enterprises to overtake others by taking a different route.

2.3 Models and Algorithms Relevant to This Study

In this section, we explore several key models and algorithms that are crucial for understanding and predicting disruptive technologies. These models include the BERT model, undirected weighted graphs, and community detection algorithms. These methods are not only innovative within their respective domains but also show great potential in predicting disruptive technologies.

BERT Model. The BERT (Bidirectional Encoder Representations from Transformers) model [18], as an advanced pre-trained language model, utilizes deep learning to capture features in language, providing new perspectives for semantic analysis of textual data. Proposed by Google in 2018 based on the Transformer architecture, the BERT model learns language representations primarily through two pre-training tasks: Masked Language Model (MLM) and Next Sentence Prediction (NSP). The MLM task enables BERT to capture semantic information and grammatical roles of words by masking some words in the input sentences and inferring them from the context. The NSP task teaches BERT to understand the logical and cohesive relationships between sentences by determining if two sentences are consecutive.

The application of the BERT model in predicting disruptive technologies is still in the exploratory stage but has already shown tremendous potential. Shi Huanting classifies the SAO structure sequences represented in patent texts using the BERT model, identifying potential disruptive technologies by analyzing the categorization of SAO structures, technological mutations, and technological integration [19]. As the model scale increases and training strategies improve, the BERT model is expected to play an increasingly important role in future technology forecasting tasks.

Undirected Weighted Graph. Undirected weighted graphs are models in graph theory, serving as powerful data structures for in-depth analysis and representation of relationships between technologies. With the exponential development of data information in recent years, graph structures have proven effective in representing data relationships, drawing widespread attention from researchers. Yang Ning and others effectively identify and cluster entities with similar features by constructing undirected weighted graphs [20]; Zhang Feng proposes a network topology vulnerability analysis method to identify vulnerable nodes in the network by using the maximum possible path of undirected weighted graphs as the collaborative production network efficiency function [21]; Liao Shunhe addresses the distance retrieval problem in undirected weighted graphs with a method based on distance signatures [22]. These studies suggest that undirected weighted graphs may be an excellent avenue for predicting disruptive technologies, helping to reveal the integration and interpenetration of knowledge across different technological fields and to identify innovation points that may trigger market changes in advance.

Community Detection Algorithm. Community detection algorithms are essential tools in graph theory for identifying community structures within graphs. This section analyzes the partitioning capabilities of community detection algorithms in undirected weighted graphs and how they can help us recognize potential technological fields and community structures. To identify potential relationships between patent vector nodes, technological fields, and community structures in undirected weighted graphs, partitioning methods such as community detection algorithms, clustering algorithms, and graph neural network-based partitioning methods are required.

Clustering algorithms, traditional partitioning methods like the K-means algorithm, do not consider the structural information of data during the calculation process and cannot fully utilize the weight information in undirected weighted graphs. Zeng Ruming points out in his article that the traditional K-means algorithm's random selection of initial cluster centers can lead to unstable clustering results [23]; Guo Yongkun notes that the traditional K-means algorithm is not ideal when dealing with datasets with significant density differences [24]. Graph Convolutional Networks (GCN), a popular neural network algorithm, can learn low-dimensional embedded representations of nodes, suitable for processing complex graph data like patent undirected weighted graphs with intricate structures and attributes. However, training GCNs requires substantial computational resources and involves adjusting many hyperparameters, making it relatively challenging to handle; Community detection algorithms, specifically designed for graph-structured data, can accurately determine community structures through modularity optimization. For example, Xiao Jing and Zhang Yongjian use community detection algorithms to enhance the understanding of community structures in complex networks, improving the division and detection capabilities of community structures through algorithm optimization [26], and Fu Lidong, Wu Hongfei, and others also apply community detection algorithms to community division in complex networks [27].

Based on the above analysis, considering the convenience, accuracy, and strong operability of community detection algorithms in partitioning undirected weighted graphs, this paper chooses to use community detection methods for community structure division.

3   Method

Section 3 will provide a detailed introduction to the prediction method for disruptive technologies based on the BERT model and graph theory techniques. First, a patent dataset is constructed and its accuracy is ensured through data preprocessing. Then, the data is structured into an undirected weighted graph, and domain division is carried out using community detection technology. By extracting cross-domain features, we identify key technological characteristics. Finally, the potential of disruptive technologies is quantitatively assessed by applying a comprehensive scoring formula for disruptive technology potential, and potential disruptive technologies are predicted through evaluation by domain experts.

3.1   Methods for Constructing Patent Datasets

When it comes to accessing patent data, users can either retrieve it directly from official patent offices, including the United States Patent and Trademark Office (USPTO), the European Patent Office (EPO), and the China National Intellectual Property Administration (CNIPA), or opt for third-party patent search platforms such as Google Patents, Derwent Innovations Index, and PatSnap. Since third-party platforms offer the convenience of aggregating data from various patent collections, thus providing a broader coverage, as well as the flexibility to precisely customize the export scope, format, and content (including titles, abstracts, and specifications), this paper ultimately selects PatSnap as the preferred platform for downloading patent text data. Figure 1 is the framework for constructing a patent dataset.

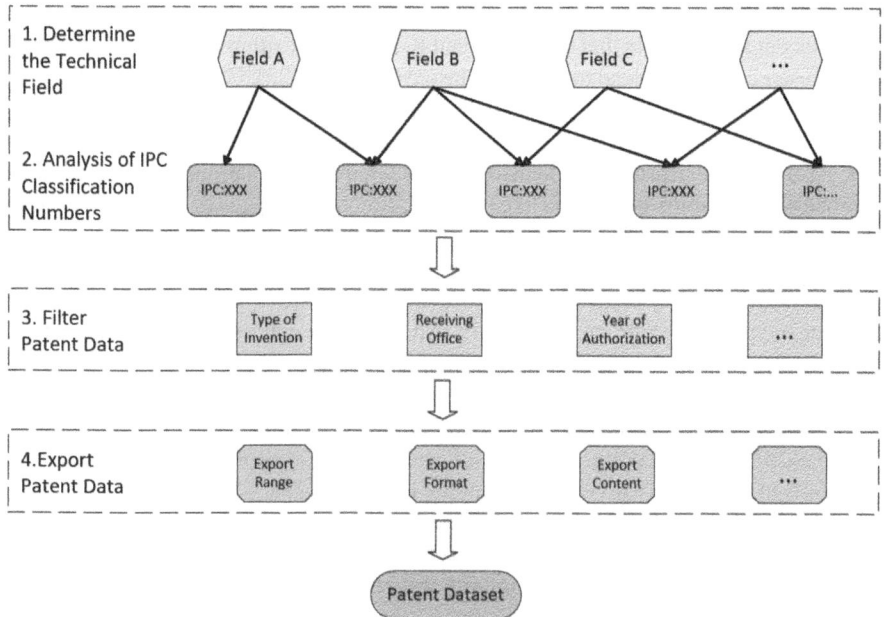

Fig. 1. Patent Dataset Construction Framework

Once the patent dataset is constructed, preprocessing operations must be performed on the patent text data, which specifically includes steps such as data cleaning and language processing. Data cleaning is a crucial step in text data preprocessing, primarily aimed at removing duplicate patent records, excluding incomplete entries, standardizing data formats, and other operations to prepare for subsequent text content processing. Subsequently, language processing is carried out on the cleaned patent data to reduce data noise and retain key information in the text.

3.2   Constructing an Undirected Weighted Graph

The construction of an undirected weighted graph mainly consists of three parts: the creation of the node dataset, the creation of the edge dataset, and the drawing of the undirected weighted graph. All these steps are implemented using the Python language in the PyCharm development environment. PyCharm is a powerful Integrated Development Environment (IDE) that provides convenient development tools for operations such as editing, debugging, and testing. The BERT model we have chosen is the bert-base-chinese model, which is a BERT model specifically pre-trained for Chinese corpora. The following will introduce the process of building the node dataset, edge dataset, and drawing the undirected weighted graph.

1. Node Dataset Construction: Commence by integrating the Tokenizer module and the Model repository from the BERT architecture into the script, and subsequently load the pertinent BERT model. Extract vectors from patent abstracts, where a selection of one hundred abstracts is randomly drawn from each domain to compile the patent abstract list. Subsequently, encode and transform these abstracts into vectors using BERT, while meticulously handling any anomalies within the abstracts to culminate in a vectorized node dataset, denoted as "abstracts."

2. Edge Dataset Construction: This phase is facilitated by the BERT architecture and entails the extraction of patent keywords and the generation of a keyword co-occurrence matrix. Initiate by setting up the tokenizer and BERT model. The extraction of patent keywords is streamlined through the development of a dedicated function. Post tokenization, format conversion, and truncation of the patent abstract data, the refined information is channeled into the BERT model to acquire the final hidden layer state. Compute the weighted average embedding vectors for each token within the hidden layer, and designate the initial ten tokens from each patent abstract as keywords. Traverse the list of abstracts exhaustively to compile all keywords. Subsequently, fabricate a nested dictionary to log the frequency of co-occurrence between any pair of keywords. Once the co-occurrence matrix is populated, transmute it into a bidimensional list array.

3. Rendering the Undirected Weighted Graph: With the patent vector list and co-occurrence matrix in hand, employ the "NetworkX" library to initiate an undirected graph, denoted as G. Iterate over the patent vector list to incorporate patent nodes. Establish a threshold variable that functions as a benchmark for edge weights to curtail the proliferation of paths within the graph. For each pair of patent data, should the weight surpass the threshold, introduce an edge between the two patents,

complete with assigned weight. Conclude by employing the "Spring_layout" algorithm to orchestrate the graphical layout and by leveraging "Pyplot" for the aesthetic visualization of the graph.

3.3 Community Domain Division

To identify potential connections between patent nodes and technological fields, it is necessary to partition the existing undirected weighted graph. According to the analysis above, this task can be accomplished through community detection algorithms, which mainly include the Louvain algorithm and the Infomap algorithm for patent data analysis. The Louvain algorithm focuses on distinguishing different community structures by optimizing modularity, which is a metric that measures the proportion of a community's internal to external connections [28]. The Infomap algorithm is based on the principle of minimum entropy, aiming to find the optimal community structure by minimizing the information entropy required for a random walker's path within the community structure [29]. It can quickly partition networks, and the novelty of the algorithm lies in the fact that it does not require the number of communities to be specified in advance. Instead, it naturally discovers community structures through the optimization process [30]. In addition, the Infomap algorithm identifies community structures by minimizing the random walk time, which allows it to provide more refined and realistic community divisions in some cases [31]. Based on the comparison and analysis of the two community detection algorithms, this paper chooses the Infomap algorithm to partition the patent undirected weighted graph.

The core of the Infomap algorithm is to minimize the Map Equation to find a suitable community division. The Map Equation was proposed by Rosvall and Bergstrom in their paper [29], as shown in Eq. (1).

$$MapEquation = \sum_{i=1}^{N} \frac{P_i}{P_{stop}} \log\left(\frac{P_{stop}}{P_i}\right) \tag{1}$$

In this equation, N signifies the total node count within the network, P_i represents the likelihood of node i being accessed during a random walk, and P_{stop} denotes the probability of the walk ceasing at any given step. The equation encapsulates the concept that the amount of additional information required by a walker to keep track of their location is contingent upon the network's community divisions. Less information is needed for traversal within a community, whereas more information is demanded when moving across different communities. The Infomap algorithm iteratively refines its approach to minimize the Map Equation's value, consequently identifying the most advantageous community structure.

3.4 Definition and Extraction of Cross-Domain Features

Shi Huanting has proposed that disruptive technologies involve the intersection and integration of technologies from different fields, and the identification of disruptive technologies needs to consider the integration of technologies [19]. This indicates that the intersection and integration of different technological fields are key to forming new technological paths. We define cross-domain integration paths as potential avenues for the fusion and integration of patent technologies from different technological fields, and these avenues foretell new directions for technological development. Li Qianrui identifies disruptive technologies through patent metrics, emphasizing the importance of the technological influence of patents [32]. Combined with Shi Huanting's application of community detection algorithms in the identification of disruptive technologies [19], we believe that cross-domain spread sources are nodes with high connectivity between different technological communities, playing an important role in promoting the cross-domain dissemination of technological knowledge.

We define the cross-domain integration path and the cross-domain spread source as important characteristics of cross-domain technology. Below are the specific identification methods for these two cross-domain features: First, use the BERT vector representation of patent data to calculate the similarity of different nodes. Nodes with higher similarity are more likely to form new paths. Select high similarity paths as potential cross-domain integration paths. Then, identify the peripheral nodes of each community, especially those connecting different communities. As mentioned above, these points may be key to cross-domain connections. Measure the betweenness centrality of these nodes (nodes with high betweenness centrality can act as bridges between different communities and promote cross-domain dissemination), and select nodes with high betweenness centrality as potential cross-domain spread sources. If the nodes in the potential cross-domain integration path coincide with the potential cross-domain spread sources, then these nodes and edges, as well as the cross-community nodes and edges in the original graph, are defined as cross-domain spread sources and cross-domain integration paths.

Predicting Potential Cross-Domain Integration Paths. This paper employs the cosine similarity measure to assess the similarity between different nodes in the undirected weighted graph. Assuming the vectors of two nodes are $A = [a_1, a_2, \ldots, a_n]$ and $B = [b_1, b_2, \ldots, b_n]$, the cosine similarity between the two vectors is as shown in the formula:

$$\cos(A, B) = \frac{A \cdot B}{\|A\| \|B\|} \tag{2}$$

In the formula, $A \cdot B$ is the dot product of vectors A and B, and $\|A\|$ and $\|B\|$ are the Euclidean norms of A and B, respectively.

After constructing the similarity matrix (which includes the similarity scores for each pair of patent vector nodes), the Dijkstra algorithm (a path-finding algorithm) is used to filter and select n paths with high similarity scores, excluding existing paths, as potential cross-domain integration paths.

Identifying Potential Cross-Domain Propagation Sources. During the partitioning phase, the Infomap algorithm assigns a corresponding community label to each patent node in the undirected weighted graph. Based on these labels, a list of patent community members is constructed. All nodes within each community are traversed, and nodes that have only a single neighbor and nodes that have at least one neighbor from a different community (boundary nodes) are selected to create a dataset of boundary nodes. The Brandes algorithm is used to calculate the betweenness centrality of boundary nodes within each domain. Betweenness centrality was proposed by sociologist Linton Freeman to identify key nodes within a network [34]. The steps for calculating betweenness centrality are as follows:

For a node v in the graph, its betweenness centrality $C_B(v)$ is given by:

$$C_B(v) = \sum_{s \neq v \neq t} \frac{\sigma_{st}(v)}{\sigma_{st}} \tag{3}$$

In the formula, σ_{st} represents the number of shortest paths between node pairs s and t, and σ_{st} represents the number of those shortest paths that pass through node v.

First, initialize the node data to ensure that the betweenness centrality of every node v in the graph is 0. Traverse the node data and calculate the number of shortest paths for each pair of nodes. For each shortest path, if it passes through node v, increment the count of σ_{st}. For each node v, calculate $C_B(v)$ using the above formula and accumulate the value.

Nodes with high betweenness centrality are filtered as potential cross-domain propagation sources. These are then compared and analyzed with the potential cross-domain integration paths obtained in the previous step. Retain the overlapping nodes and edges, as well as the original inter-community nodes and edges, as the cross-domain propagation sources and cross-domain integration paths.

3.5 Disruptive Technology Forecasting

Based on the results of cross-domain feature extraction, cross-domain propagation sources and cross-domain integration paths can be identified. Utilizing these node sets and edge sets, a cross-domain undirected weighted graph is constructed. An analysis of the nodes within this graph leads to the proposal of an integrated predictive scoring formula for the potential of disruptive technologies, as shown in Eq. (5). This formula takes into account the similarity of patent pairs, betweenness centrality, and community structure characteristics. A threshold is set according to the scoring results, and patent pairs with scores above the threshold are selected as disruptive technology integration patent pairs. The selection results are then verified by experts in the respective fields, and feedback is obtained to adjust the parameters of each step.

$$PairScore = \frac{1}{2}\left(Score_{node_A} + Score_{node_B}\right) + \lambda \cdot PathScore \tag{4}$$

In the equation, λ is the weight coefficient, $Score_{node_i}$ represents the comprehensive score of patent node i, and $PathScore$ represents the comprehensive score of the path between patents A and B.

(1) Comprehensive Node Score:

$$Score_{node} = \omega_1 \cdot Similarity + \omega_2 \cdot Centrality + \omega_3 \cdot Influence \tag{5}$$

In the equation, ω_1, ω_2, ω_3 are weight parameters. *Similarity* is the similarity score between patent node pairs, calculated using cosine similarity; *Centrality* is the centrality score, obtained by calculating betweenness centrality; Influence is the community influence score, determined by combining the total number of edges, total weight between the community and other communities, and the external connections of all nodes within the community.

$$Influence = \omega \cdot Edges + \phi \cdot Weight + \psi \cdot Connectivity \qquad (6)$$

In the equation, ω, ϕ, ψ are weight parameters, and Edges, Weight, Connectivity represent the total number of edges, total weight, and community external connections, respectively.

(2) Comprehensive Path Score:

$$PathScore = \frac{UT}{TT} \times Entropy \qquad (7)$$

$$Entropy = -\sum_{i=1}^{n} P_i \log_2 P_i \qquad (8)$$

In the equations, UT (Unique Tech Fields) represents the unique number of different technology fields on the path, TT (Total Tech Fields) represents the total number of technology fields on the path, and Entropy is the technology category entropy. "n" is the total number of community domains in the graph, and P_i represents

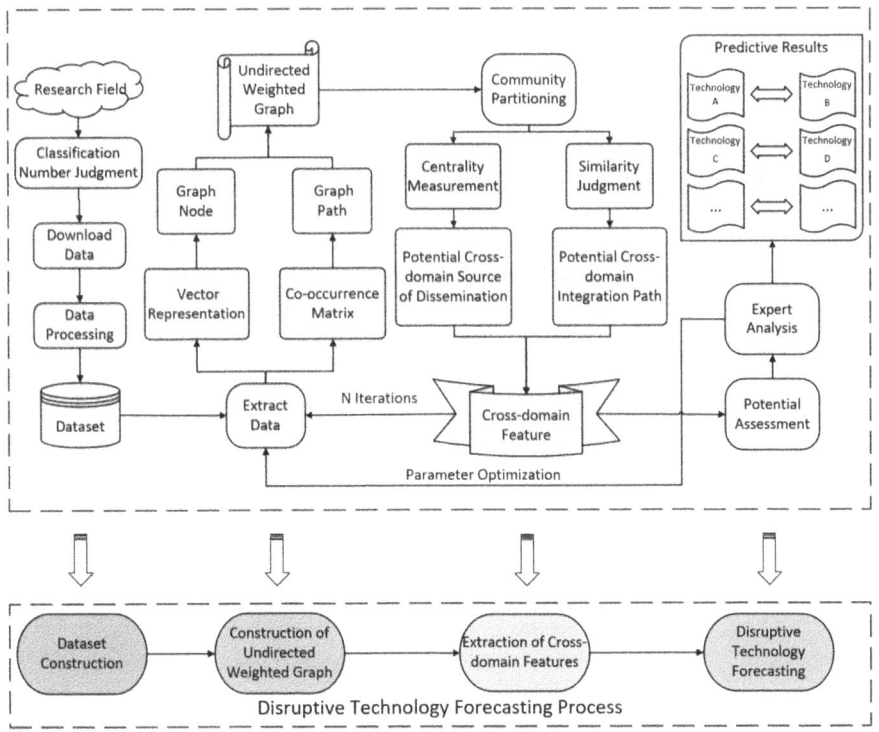

Fig. 2. Disruptive Technology Forecasting Framework

the proportion of this domain among the total domains. Entropy, introduced by Claude Shannon in his 1948 paper "A Mathematical Theory of Communication," measures the uncertainty of information [35]. Here, technology category entropy is used to quantify the distribution of different domain technologies in a path.

Based on the aforementioned integrated predictive scoring formula for disruptive technology potential, patent pairs in the cross-domain undirected weighted graph can be evaluated to identify patent integration pairs with high disruptive potential. The prediction results are analyzed by domain experts for the final judgment of future disruptive technology types. Figure 2 illustrates the overall framework for disruptive technology forecasting.

4 Predictive Process

4.1 Construction of Patent Dataset and Dataset Preprocessing

This paper focuses on several popular sub-domains within the larger categories of mechanics, chemistry, physics, and operations, which not only have high potential for innovation and interdisciplinary characteristics but also hold significant positions in current technological development. These domains are recognized as fertile grounds for the emergence of disruptive technologies. We have chosen to conduct in-depth research on sub-domains such as additive manufacturing, robotics and automation, artificial intelligence, adaptive structures, and composite material manufacturing for the following reasons:

1. Technological forefront: The selected fields are at the forefront of current technology, representing the latest developments in their respective disciplines.
2. Interdisciplinary integration: The selected fields are highly cross-disciplinary in technology, facilitating the integration of technologies across different disciplines.
3. Market potential: The selected fields have substantial market application potential, capable of driving the development of new production and business models.
4. Disruptive impact: Technological advancements in the selected fields may have a disruptive impact on existing markets and landscapes.

Table 4. List of IPC Classification Codes for Various Domains

Research Field	IPC Code
Additive Manufacturing	B29C64/00 (Rapid Prototyping, 3D Printing Technology), etc.
Robotics and Automation	B25J (Robots), F16B (Mechanical Manipulation), B64G1 (Automated Equipment), etc.
Artificial Intelligence	G06T (Image Data Processing and Generation), G06F (Computation, Information Storage), etc.
Adaptive Structures	F16H (Structural Components of Machines or Devices, particularly Adaptive Structures), etc.
Composite Material Manufacturing	C04B (Non-metal Materials), B29C (Composite Material Manufacturing), etc.

The innovative integration of these fields not only meets the conditions for the emergence of cross-domain disruptive technologies but may also give rise to new production and business models, exerting a disruptive influence on the existing market and technological landscape. After analyzing the functions and characteristics of patents in different fields, we have determined the corresponding IPC classification codes for each field as shown in Table 4.

Query the IPC classification codes mentioned above for corresponding patents, and select and download the invention patent data from each field within the past three years.

公开(公告)标题	法律状态/	[标]当前申	申请日	公开(公告)	专利文本信息
CN126183综合计算机	未缴年费	株式会社₹	######	######	生成用于高精度快速加工的数控数据的计算机辅E
CN103901计算机辅助	授权 | 权利	三菱动力权	######	######	本发明提供计算机辅助制造装置、产品形状加工;
CN100468钣金冲压	未缴年费	富士康科₹	######	######	一种钣金冲压计算机辅助制造方法,该方法包括;
CN115328一种计算机	授权	广东亚数₹	######	######	本申请提供了一种计算机辅助制造的方法、装置、
CN113536一种工业图	授权	苏州悦谱	2021-7-8	######	本发明公开了一种工业图形计算机辅助制造的信
CN113836工业图形;	授权	苏州悦谱	######	######	本发明涉及工业图形计算机辅助制造信号线路数
CN115774G代码的生	授权	济南邦德	######	######	本申请属于程序控制技术领域,具体涉及一种G(
CN113435一种工业图	授权	苏州悦谱	######	######	本发明公开了一种工业图形计算机辅助制造的防;
CN105358端铣削设备	未缴年费	国立大学;	######	######	本发明解决提供端铣削设备以及计算机辅助制造(
CN113254一种基于]	授权	清华大学 |	2021-2-7	######	一种基于工业PAAS构建的协同制造引擎系统,该
CN109152用于牙科更	授权	西诺德牙₹	######	######	本发明涉及一种牙科更换部件或牙科辅助元件由;
CN10435C机器的CAľ	授权 | 权利	赫克斯冈:	######	2018-2-2	用于计算机数控(CNC)设备的系统和方法一个包括
CN103425一种自由	未缴年费	大连理工.	######	2016-7-6	一种自由曲面灰度掩膜的设计方法,属于微制造;
CN100363假肢接受腔	未缴年费	北京航空	2004-6-9	######	本发明公开了一种假肢接受腔快速成型装置,包;
CN102183开槽机数排	授权 | 权利	上海维宏	2011-3-1	2013-1-2	本发明涉及一种开槽机数控系统中实现开槽位置;
CN103237假牙的建档	授权	3形状股份	######	######	公开了用于建模和制造患者假牙的方法,其中该
CN108549一种自由	授权	合肥工业;	######	######	本发明涉及一种自由曲面弧长参数曲线加工轨迹;
CN113435印刷电路机	授权	苏州悦谱	2021-7-8	######	本发明公开了一种印刷电路板铜面数据间距调整
CN111819用于工具的	授权 | 权利	赫克斯冈:	######	######	用于在计算机辅助制造(CAM)系统(100)中确定优(

Fig. 3. Patent Dataset

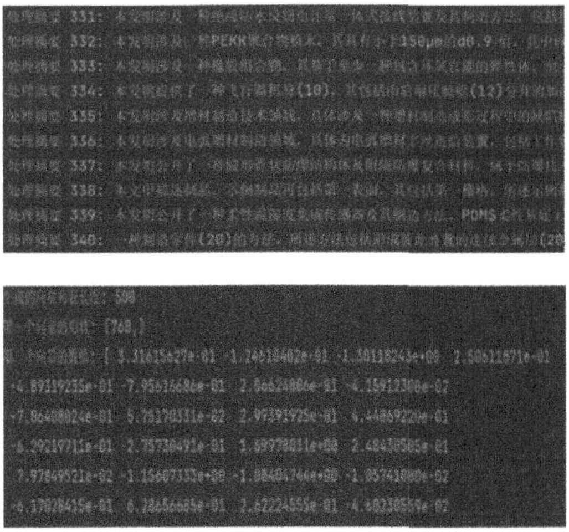

Fig. 4. Partial Patent Preprocessing Results and Vectorization Results

The dataset is shown in Fig. 3. Clean and process the downloaded patent dataset linguistically, and use the BERT model to vectorize the patent data. Some of the processing results and vector representations are shown in Fig. 4.

4.2 Constructing an Undirected Weighted Graph

For the processed list of patent data, encoding and vector operations are performed through the BERT model to obtain a vectorized representation of the node dataset list. The top ten keywords from each patent data are extracted, a keyword co-occurrence matrix is constructed and visually processed, as shown in Fig. 5.

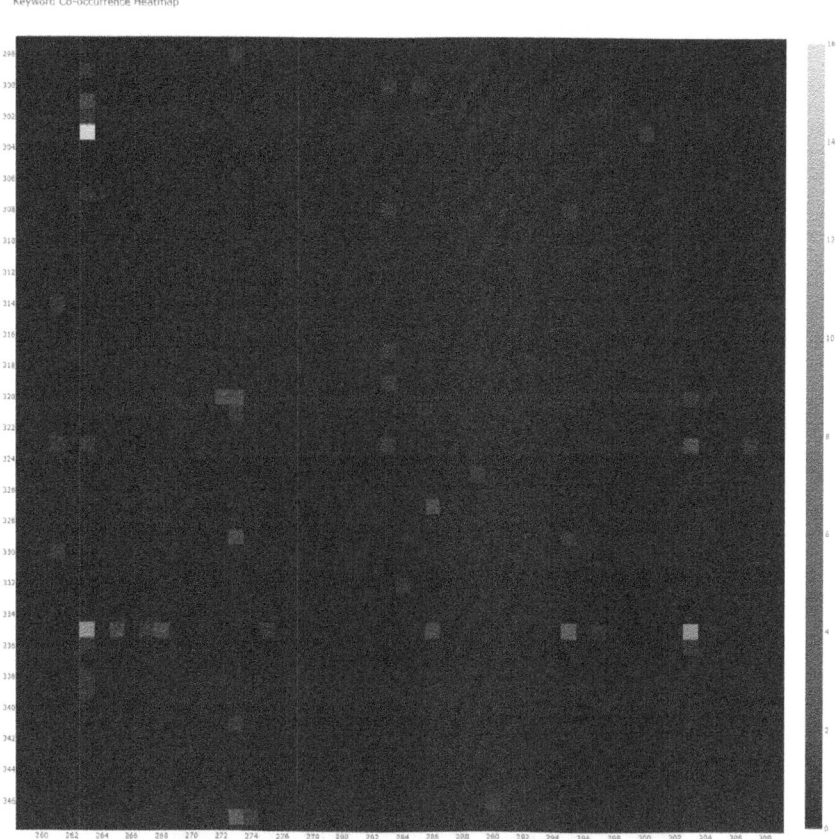

Fig. 5. Visualization of Keyword Co-occurrence Matrix

Plotly is used to create a heatmap of the co-occurrence matrix, where the more times a pair of keywords appears, the brighter the color of the corresponding spot, indicating the higher frequency of the two keywords appearing in the papers and the greater role they play in the patent data.

By creating an undirected weighted graph with the constructed patent keyword co-occurrence matrix and vector list, and filtering for edges with a co-occurrence frequency greater than 3, the Matplotlib library is used to draw the undirected weighted graph. The Plt.show() function is called to display the entire graph, with the results shown in Fig. 6.

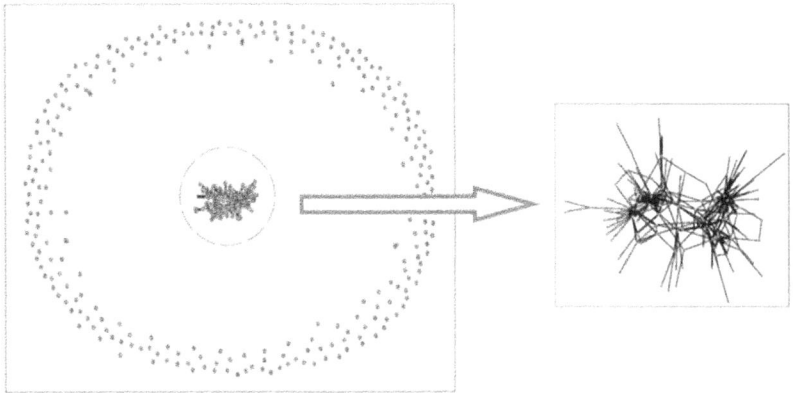

Fig. 6. Patent Dataset Undirected Weighted Graph

4.3 Domain Division

Based on the vector list and co-occurrence matrix, nodes and edges are created in Infomap, and weight parameters are set for the edges. Then, the Infomap algorithm

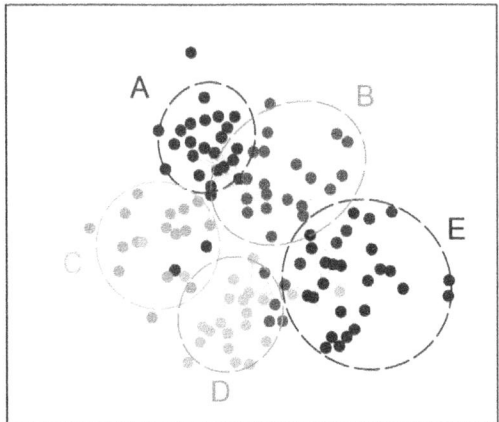

Fig. 7. Community Domain Division Diagram

is applied to obtain community divisions. Utilizing NetworkX and Matplotlib, the community structure is visually processed. The community partition structure is shown in Fig. 7. According to the partition results, five distinct blocks of clusters A, B, C, D, and E can be identified. Edge nodes within each community are more easily distinguishable, and these nodes are more receptive to diverse information and knowledge, possessing a higher potential in promoting cross-domain technology integration and the emergence of disruptive technologies.

4.4 Extracting Domain Features and Predicting Disruptive Technologies

The cosine similarity formula is used to calculate the similarity between each pair of patents, constructing a cosine similarity matrix that includes the cosine similarity values for each pair of patent nodes. Patent pairs with similarities greater than a specified threshold are selected as potential cross-domain integration paths. Utilizing the community structure identified by the Infomap algorithm, edge nodes of the communities are located, and the betweenness centrality of these edge nodes is calculated to determine which nodes can act as bridges between different communities. By combining patents with high cosine similarity and patents with high betweenness centrality, cross-domain propagation sources and cross-domain integration paths are selected and visually processed, as shown in Fig. 8.

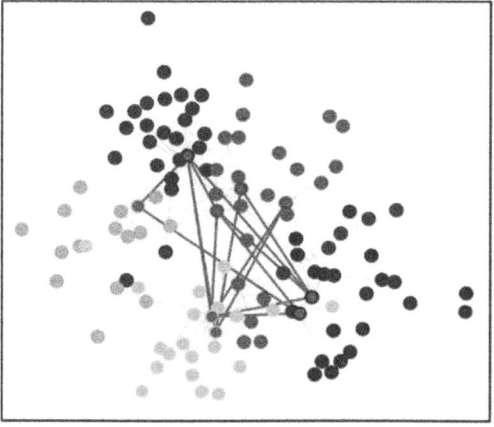

Fig. 8. Cross-domain Feature Extraction

Considering that the program randomly selects one hundred patents from each field each time, the results are subject to randomness and uncertainty. Therefore, the aforementioned process is subjected to iterative processing. The outcomes of each iteration are imported into the Comprehensive Disruptive Potential Scoring Formula, and a threshold for disruptive potential is set. Patent pairs with a potential greater than the disruptive potential threshold are selected as Disruptive Technology Potential Integration Patent Pairs. Domain experts then screen and judge these potential integration patent pairs to infer which technologies may converge and disrupt the existing market in the future. The expert assessment results are shown in Table 5, and the prediction process is illustrated in Fig. 9.

Table 5. List of IPC Classification Codes for Various Domains

No.	CD	OTM	TI	IP	Ass
001	Yes	High	Significant	High	A
002	Yes	Medium	Limited	Low	C
…	…	…	…	…	…
00N	No	Medium	Limited	Medium	B

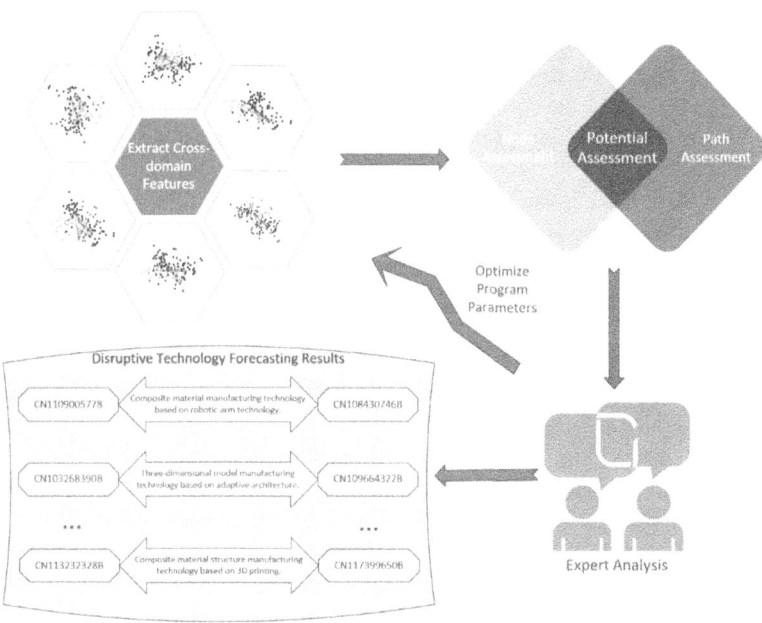

Fig. 9. Predicting Disruptive Technologies Based on Cross-domain Feature Data

Domain experts conduct a comprehensive evaluation of the disruptive potential following the integration of two technologies across four critical metrics: CD (cross-domain nature), OTM (maturity of the original technology), TI (level of technological innovation) and IP (potential for technological integration).

Initially, they assess whether the technologies span across domains, as intra-domain amalgamations are less likely to spawn disruptive innovations; The maturity of the original technology is stratified into high, medium, and low, providing a gauge to determine the technology's readiness for widespread commercial application. Technological innovation is classified into significant, moderate, and limited, which aids in evaluating the technology's capacity to exert a disruptive influence on established or mainstream technologies; The potential for technological integration is categorized into high, medium, and low, which is pivotal for assessing the feasibility of merging the two technologies. By quantifying the scores of these four metrics and aggregating them with their respective weights, a composite assessment is derived; The Ass(composite assessment) is tiered into A, B, and C levels, with 'A' representing the highest disruptive potential and 'C' indicating the lowest.

4.5 Analysis of Forecast Results

After expert analysis and judgment, a variety of technologies that may integrate and disrupt the existing market in the future have been identified. The following will list some of the integration technologies with strong disruptive potential as examples, and conduct in-depth analysis on them from three directions: innovation, market application potential, and future research value.

(1) Integration of robotic arm operation methods with the preparation methods of fiber composite materials, the combination of technologies in the "Robotics and Automation" field with the "Composite Material Manufacturing" field, has the following advantages:

Innovation: The precise motion control capability of the robotic arm can greatly improve the laying accuracy of composite materials, ensuring the integrity and consistency of material properties; it reduces the need for manual operation, decreases human error, and increases production efficiency; it can adapt to complex molds and multidimensional surfaces, providing a foundation for manufacturing complex-shaped composite material structures.

Market Application Potential: In the aerospace field, it can be used to manufacture lightweight and high-strength aircraft parts; in the automotive industry, it can be used to produce lightweight car parts, reducing production costs; in the sports equipment field, it offers the possibility of manufacturing customized and optimized design equipment such as high-performance bicycles and tennis rackets.

Future Research Value: Research on the compatibility of different types of composite materials with robotic arm operations, optimization of laying techniques and parameters; development of more intelligent control systems to achieve self-adaptive adjustment and self-optimization of robotic arm operations; evaluation of economic benefits after scaling up, exploring ways to reduce costs.

(2) Integration of 3D printing technology with the manufacturing of curved building module components, the combination of technologies in the "Additive Manufacturing" field with the "Adaptive Structures" field, has the following advantages:

Innovation: Based on the unlimited design freedom of 3D printing technology, it is possible to create components of any shape and structure; it can significantly speed up construction and shorten project completion times; 3D printing can reduce the generation of construction waste by precisely controlling material placement, improving material utilization.

Market Application Potential: It can be used to quickly manufacture building modules and complex curved structures, reducing construction costs; it can provide architects with the possibility of designing more aesthetically pleasing and personalized designs.

Future Research Value: Improve the structural integrity and durability of 3D printed building modules to meet construction standards; assess the environmental impact of 3D printing technology; develop integrated 3D printing construction systems to integrate design, manufacturing, and construction processes.

(3) Artificial intelligence-assisted additive manufacturing technology, the combination of technologies in the "Additive Manufacturing" field, "Artificial Intelligence" field, and "Adaptive Structures" field, has the following advantages:

Innovation: AI can assist in design, optimizing 3D printing models to achieve lighter, lower-cost, and better-performing products; combined with AI adaptive manufacturing systems can adjust the production process according to real-time data, meeting the needs of different types of products.

Market Application Potential: In the manufacturing industry, AI-assisted 3D printing technology can be used for rapid manufacturing and production; for customized production of medical devices to improve treatment effects; in the aerospace field, for manufacturing complex lightweight parts to enhance product performance.

Future Research Value: Further development and optimization of AI algorithms to improve decision-making capabilities in the design process; exploration of the intersection of AI, 3D printing, and material science to promote interdisciplinary integrated solutions.

5   Conclusion

This paper proposes a novel model for predicting disruptive technologies, integrating various scientific methods such as the BERT model and graph theory, offering a new perspective for forecasting and identifying cross-disciplinary integrable disruptive technologies. The main contributions of this study are summarized as follows:

1. Construction of Interdisciplinary Methodology: This study combines the text vectorization technology of the BERT model with graph analysis techniques to establish a new data processing and analysis framework, providing a convenient and practical new method for the prediction and identification of disruptive technologies.
2. In-depth Mining of Patent Datasets: This study selects and constructs patent datasets in popular technological fields, using keyword co-occurrence matrices to draw undirected weighted graphs, laying the foundation for the visualization analysis of technologies in different fields.
3. In-depth Integration of Multiple Disruptive Technology Forecasting Methods: This study combines the advantages of patent analysis, bibliometrics, and expert systems, deeply mining patent datasets while considering the content of patent technologies and the judgments of domain experts, enabling comprehensive forecasting and identification of disruptive technologies.

4. Extraction and Analysis of Cross-Domain Technological Features: By analyzing cross-domain patent data, this study captures potential connections and integration trends between cutting-edge technologies in different fields. It also proposes a comprehensive scoring formula for the disruptive potential of patent technology integration, as well as the concepts of cross-domain spread sources and cross-domain integration paths, which help identify disruptive technologies that may have a significant impact across multiple domains.

5. Application and Innovation of TRIZ Theory: This study innovatively applies the Ideal Solution Analysis and Resource Analysis from TRIZ theory to the construction and optimization of the disruptive technology forecasting model. By conducting an in-depth Ideal Solution Analysis of existing prediction methods, we identified the strengths and limitations of each method and used Resource Analysis to systematically mine and utilize internal and external resources, including patent data, interdisciplinary knowledge, and expert insights, to enhance the comprehensiveness and accuracy of the forecasting model.

This study still has some limitations and deficiencies. Firstly, this paper only mines patent data and does not consider the impact of market, papers, and other data, which may lead to inaccurate analysis results. Secondly, although the BERT model belongs to the field of deep learning and provides strong analytical capabilities for text mining, the current study's integration method of the BERT model's Tokenizer module and Model repository is quite basic, and further optimization is needed in keyword extraction and context-aware processing. In addition, due to the subjective influence of domain experts, the final results may have some bias. In subsequent work, we will further delve into the following aspects:

1. Multi-source Data Fusion: In the future, we will integrate multi-dimensional data such as market reports, academic papers, and news information to provide a more comprehensive perspective for predicting disruptive technologies and improve the accuracy of predictions.

2. Optimization of the BERT Model: Optimize the integration of the BERT model and keyword extraction technology, enhance context-aware processing capabilities, and improve data preprocessing procedures to ensure the completeness and accuracy of information.

3. Enhancement of Graph Theory Techniques: Develop new graph theory algorithms or improve existing ones to more finely identify community structures and key nodes, enhancing predictive capabilities.

References

1. Su, P., Su, C., Pan, Y.T.: The current status and enlightenment of disruptive technology identification methods. Libr. Inf. Serv. **63**(20), 129–138 (2019)
2. Wang, Z.Y., Dang, X.L., Liu, C.H., et al.: The basic characteristics of disruptive technology and the main practices of foreign research. Natl. Defense Sci. Technol. **36**(3), 14–17 (2015)
3. Wang, A., Xu, Y., Sun, Z.T., et al.: A brief analysis of foreign disruptive technology identification methods. Chin. Eng. Sci. **9**(5), 79–84 (2017)

4. Xing, X.Z., Ren, L., Lei, X.P., et al.: Research on the identification of disruptive technology based on the evolution of patent themes: taking the brain-like intelligence field as an example. Inf. Sci. **41**(03), 81–88 (2023)
5. Funk, R.J., Owen-Smith, J.: A dynamic network measure of technological change. Manag. Sci. **63**(3), 791–817 (2017)
6. Cheng, Y., Huang, L., Ramlogan, R., Li, X.: Forecasting of potential impacts of disruptive technology in promising technological areas: elaborating the SIRS epidemic model in RFID technology. Technol. Forecast. Soc. Change **117**, 170–183 (2017)
7. Liu, H., Zhang, S.C., Li, L.Y., et al.: Research on the identification of disruptive technology based on bibliometrics and measurement of disruptive potential: taking 5G technology as an example. Sci. Technol. Ind. **23**(06), 16–25 (2023)
8. Wang, C., Ma, M., Wang, H.Y., et al.: Research on the diffusion characteristics of disruptive technology from the perspective of life cycle. J. Inf. Sci. **41**(08), 845–859 (2022)
9. Li, R., Yu, W., Huang, Q., Chen, Q.: A new identify disruptive technologies algorithm based on technology develop network. Math. Probl. Eng. **2022**(1), 7354535 (2022)
10. Huang, Q., Guo, Y., Jiang, J., Ming, B.: China's clean power development path under the goal of "dual carbon." J. Shanghai Jiao Tong Univ. **55**, 1499–1509 (2021)
11. Cheng, G., Peng, Y., Huang, J., et al.: In-depth analysis of the ideal final result (IFR) in TRIZ theory. Sci-Tech Innov. Brands **23**(06), 63–66 (2023)
12. Tan, R.H.: TRIZ and Its Application - The Process and Methods of Technological Innovation. Higher Education Press, Beijing (2010)
13. Altshuller, G.S.: Creativity as an Exact Science: The Theory of the Solution of Inventive Problems. Gordon and Breach, New York (1984)
14. Huang, C.H., Yin, J., Hou, F.A.: Text similarity measurement method combining lexical semantic information and TF-IDF method. J. Comput. Sci. **34**, 856–864 (2011)
15. Zhang, J.: Research on the Construction and Verification of the Identification Index System of Disruptive Technology. University of Chinese Academy of Sciences (2022)
16. Wang, Y.W., Zhou, Y., Chen, L.Y.: Research on the evolution path of disruptive technological innovation in china based on patent analysis—taking nanotechnology innovation as an example. Sci. Technol. Manag. Res. **40**(1), 124–131 (2020)
17. Zhang, J.W., Dong, Y.: Research progress on the identification indicators of disruptive technology. Theory Pract. Inf. **43**(06), 194–199+193 (2020)
18. Garud, R., Karnøe, B.: Bricolage versus breakthrough: distributed and embedded agency in technology entrepreneurship. J. Bus. Ventur. **18**(5), 599–620 (2003)
19. Markard, J., Truffer, B.: Innovation processes in large technical systems: market liberalization and the case of the European electricity sector. Res. Policy **37**(4), 635–651 (2008)
20. Cammarano, A., Caputo, M., Lamberti, E., Michelino, F.: Open innovation and intellectual property: a knowledge-based approach. Manag. Decis. **55**(6), 1182–1208 (2017)
21. Devlin, J., Chang, M.W., Lee, K., et al.: BERT: pre-training of deep bidirectional transformers for language understanding. In: Proceedings of the 2019 Annual Conference of the North American Chapter of the Association for Computational Linguistics: Human Language Technologies, Minneapolis, USA, pp. 4171–4186 (2019)
22. Shi, H.T.: Research on the Identification Method of Disruptive Technology Based on Deep Learning Model. North University of China (2023)
23. Yang, N., Lu, J., Shao, Q., et al.: Research on the entity recognition method of unmodeled based on undirected block weighted graph. Comput. Appl. Res. **38**(01), 169–174 (2021)
24. Zhang, F., Yang, Y., Jia, J.G., et al.: Research on the vulnerability analysis method of collaborative production network based on undirected weighted graph. China Mech. Eng. **23**(10), 1216–1220 (2012)
25. Liao, S.H., Le, J.J.: A distance retrieval and query method based on undirected weighted graph. Comput. Eng. **43**(15), 80–82 (2008)

26. Zeng, R.M.: Research on the Improvement and Application of K-Means Clustering Algorithm. China West Normal University (2023)
27. Guo, Y.K., Zhang, X.Y., Liu, L.P., et al.: K-means clustering algorithm with optimized initial cluster centers. Comput. Eng. Appl. **56**(15), 172–178 (2020)
28. Xin, H.L., Ding, Y.Q., Gao, M., et al.: Skeletal action recognition based on multi-partition spatiotemporal graph convolutional network. Sig. Process. **38**(02), 241–249 (2022)
29. Xiao, J., Zhang, Y.J., Xu, X.K.: Research progress on fuzzy overlapping community detection in complex networks. Complex Syst. Complex. Sci. **14**(03), 8–29 (2017)
30. Fu, L.D., Wu, H.F.: Community detection in complex networks based on enhanced similarity and Louvain method. Inf. Technol. **23**(10), 12–16 (2023)
31. Schaub, M.T., Jannesari, A., Lusher, D.: Efficient Louvain: a fast implementation of the Louvain method for community detection in large networks (2017)
32. Rosvall, M., Bergstrom, C.T.: Maps of random walks on complex networks reveal community structure. Proc. Natl. Acad. Sci. **105**(4), 1118–1123 (2008)
33. Lancichinetti, A., Fortunato, S.: Community detection algorithms: a comparison. Phys. Rev. E **80**(5), 056117 (2009)
34. Zhu, Y.P.: Research on the Classification of Interdisciplinary Science Journals Based on Community Discovery Algorithms. Tianjin Normal University (2022)
35. Li, Q.R., Guo, J.F., Huang, Y., et al.: Research on the evolution of disruptive technology themes from the perspective of mutation and fusion. Stud. Sci. Sci. **39**(12), 2129–2139 (2021)
36. Salton, G., Wong, A., Yang, C.S.: A vector space model for automatic indexing. Commun. ACM **18**(11), 613–620 (1975)
37. Freeman, L.C.: A set of measures of centrality based on betweenness. Sociometry **40**(1), 35–41 (1977)
38. Shannon, C.E.: A mathematical theory of communication. Bell Syst. Tech. J. **27**(3), 379–423 (1948)

Research on the Identification and Analysis of Technological Opportunities Utilizing the BERT Model and MULTIMOORA Approach

Jianguang Sun[1,2(✉)], Runze Miao[1,2(✉)], YuJuan Du[3], and Delong Zhang[1,2]

[1] School of Mechanical Engineering, Hebei University of Technology, Tianjin 300401, China
sjg@hebut.edu.cn, mrz202009@163.com
[2] National Engineering Research Center for Technology Innovation Method and Tool, Hebei University of Technology, Tianjin 300401, China
[3] China North Engine Research Institute, Tianjin 300401, China

Abstract. Product innovation is considered a crucial component of enhancing corporate competitiveness. Identifying technological opportunities plays a pivotal role in the success of a company's research and development (R&D) activities. Technological opportunities are defined as the potential for technological advancements in specific areas. Historically, product innovation often relied on identifying singular technological opportunities. However, the current challenge lies in recognizing and leveraging a series of interacting multiple technological opportunities. To address this, our study introduces a novel approach based on the Transformer-based Bidirectional Encoder Representations from Transformers (BERT) model. This method transforms multiple technological objectives into generalized functional behaviors by restructuring them. An in-depth analysis of patent databases is conducted using the semantic search tool from Patsnap, extracting patent information related to these functional requirements. Subsequently, the data undergoes a transition from qualitative to quantitative analysis using spherical fuzzy sets. Finally, the quantified technologies are ranked for opportunities using the MULTIMOORA method, thus completing the identification of single or multiple interacting technological opportunities.

Keywords: TRIZ · BERT Model · MULTIMOORA · Deep Learning · Technology Opportunity Analysis

1 Introduction

Technological innovation is essential for scientific progress and commercial development within industries, providing a foundation for sustained competitive advantage [1, 2]. Rapid identification and exploitation of technological opportunities are key to enhancing a company's competitiveness [3, 4]. Researchers can effectively pinpoint potential

© IFIP International Federation for Information Processing 2025
Published by Springer Nature Switzerland AG 2025
D. Cavallucci et al. (Eds.): TFC 2024, IFIP AICT 735, pp. 131–146, 2025.
https://doi.org/10.1007/978-3-031-75919-2_8

technological opportunities by analyzing technology hotspots and patent citations [5–8]. Technological opportunities can be categorized into three identification pathways: blank technology, emerging technology, and converging technology [9]. Cho visualized and mapped patent gaps within technological domains to identify opportunities; however, the resulting technological opportunities could not always be fully explained [10]. Yun utilized patent data to identify evolutionary patterns based on the effects and objectives of technologies, thereby identifying emerging technological opportunities [11]. Additionally, evolutionary patterns from other industries were referenced as a method for companies to discover their own technological opportunities. Pershina and colleagues identify cross-disciplinary technological opportunities using a knowledge combination model [12]. The recombination of knowledge systems, through fuzzy analysis of technologies for compatibility or complementarity, facilitates the permeation and recombination between technologies, thus completing the identification of technological opportunities. However, this approach is limited to the identification of single technological opportunities. As technologies become more complex and market competition intensifies, product design often involves the identification of multiple technological opportunities. As shown in Fig. 1, as the number of technological objectives increases, the complexity of identifying technological opportunities and the reliability between different technological opportunities become challenging to compute.

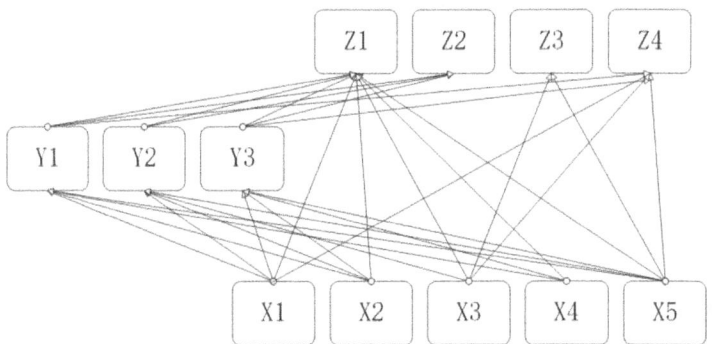

Fig. 1. Identification of Complex Technological Opportunities

Artificial Intelligence (AI), as a cutting-edge technology, is particularly noted for its potential in capturing technological opportunities and facilitating technological integration. The application of AI primarily focuses on predicting technological trends, identifying potential opportunities for technological integration, and optimizing the innovation process. The introduction of AI not only enhances the accuracy of predictions but also enables the handling of large-scale unstructured data [13]. Ajay utilized network embedding methods, fuzzy models, and deep learning models to predict potential future links within technological networks. Experiments were conducted on temporal, bidirectional, and heterogeneous networks for comparison. It was found that local and quasi-local methods performed well, while global methods tended to increase noise within the network [14]. Techniques such as Dynamic Topic Modeling (DTM) are employed to track the evolution of technology themes, which is crucial for understanding how technologies merge and diverge over time [15]. However, within a cross-disciplinary context, the use of technological terminology often suffers from semantic ambiguity and generally lacks the necessary contextual information.

This paper aims to address the following issues: First, the utilization of a trained BERT model to more accurately and efficiently process complex technical literature and patent data. Second, the integration of the Wisdom Bud platform to construct a technological knowledge flow network, enhancing the efficiency of identifying technological opportunities. Finally, the application of spherical fuzzy sets and the MULTIMOORA method to manage the uncertainty and ambiguity associated with multiple interacting technological opportunities.

2 Proposed Method

We will discuss the methodology adopted to establish the proposed framework. Figure 2 provides a global overview. To facilitate a better understanding of the steps taken, we will explore them individually.

2.1 Transforming Technological Objectives into Functional Behaviors

Identifying Technological Objectives in Target Domains
In engineering design, the technological characteristics of a product define its functions. These functions are intricately linked to specific technical details while also being influenced by abstract technological concepts. Thus, by analyzing specific problems and technological solutions, we can connect these practical applications with abstract knowledge, thereby effectively uncovering functional requirements in the target domain.

Transforming Technological Objectives into Functional Behaviors
In this research, we utilized the Transformer-based BERT model to accurately parse and process complex textual data during the transformation of engineering objectives. The BERT model's internal mechanism comprises several key components:

Initially, the embedding layer transforms each word into a vector representation, which includes lexical embeddings, segment embeddings, and position embeddings. These embeddings collectively facilitate the comprehensive vectorial representation of each word. Subsequently, these vectors are processed through multiple layers of bidirectional Transformer encoders. Utilizing a self-attention mechanism, the model captures global semantic information and contextual features, resulting in a synthesized vector encoding. The synthesized encodings are further refined through several Transformer blocks within the hidden layers. Each block is designed to enhance the model's ability to interpret and utilize the nuances in the text effectively. Ultimately, the output layer produces synthesized vector encodings for downstream tasks. These encodings convert textual descriptions of technological objectives—$\{T_1, T_2, \ldots, T_n\}$—into high-dimensional feature vectors vi. In this configuration, h_{ij} represents the vector of the j-th word in the i-th text at the hidden layer level, L denotes the total number of words in text T_i, α_{ij} is the weight determined by the attention mechanism, and LayerNorm refers to the layer normalization process that standardizes the vector outputs across different

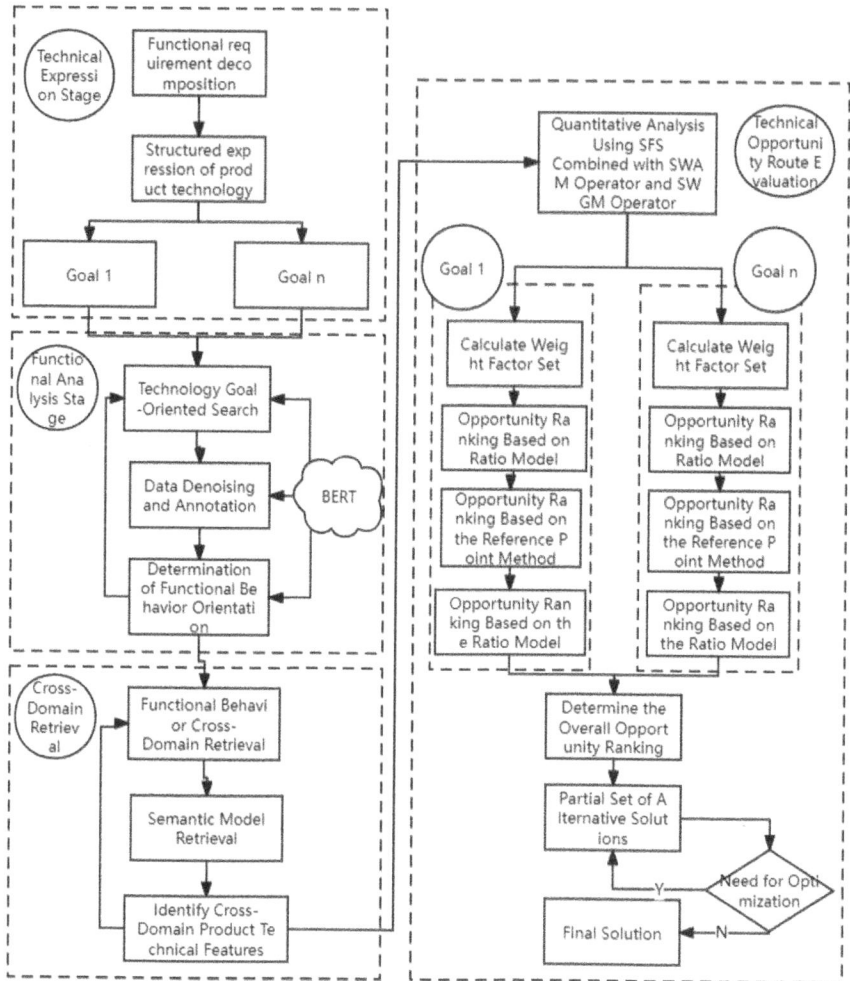

Fig. 2. TOA Process Flowchart

layers to ensure consistency and enhance the model's training efficiency. As shown in Eq. 1.

$$\mathbf{v}_i = \text{BERT}_{\text{encode}}(T_i) = \text{LayerNorm}\left(\sum_{j=1}^{L} \alpha_{ij}\mathbf{h}_{ij}\right) \tag{1}$$

The Masked Language Model (MLM) training method is completed by maximizing the log-likelihood of the masked tokens using the following formula. Let w_1, w_2, w represent the sequence of tokens in the input text, with the model's task being to predict the randomly masked token w_m. In practice, approximately 15% of the tokens are randomly masked and replaced with a special [MASK] token. The output vector for each masked position is computed through the model's multi-layer Transformer architecture, and these output vectors are then passed through a softmax layer to predict the original tokens. As shown in Eq. 2. The objective of the model is to adjust the network parameters to maximize the following:

$$L = \sum_{m=1}^{M} \log P(w_m | w_1, ..., w_{m-1}, [MASK], w_{m+1}, ..., w_N) \tag{2}$$

To clearly delineate the connections between technological objectives and functional behaviors, and to uncover the deep semantic relationships between textual data and technological knowledge, it is necessary to perform operations as indicated by Eq. 3. In this operation, each feature vector vi is transformed into a knowledge abstraction Ki through a predefined nonlinear transformation function ϕ_k and corresponding weights β_k. This transformation allows for the extraction and refinement of underlying technological insights from complex textual datasets, facilitating a more nuanced understanding of how textual descriptions align with practical technological applications.

$$K_i = g\left(\sum_{k=1}^{m} \beta_k \phi_k(\mathbf{v}_i)\right) \tag{3}$$

Cross-Domain Retrieval of Functional Behaviors

Patsnap leverages its fine-grained capabilities in entity linking, problem association, domain correlation, and feature selection to effectively govern data. Coupled with its proprietary machine translation engine, Patsnap effectively eliminates language barriers that exist between user semantics and intent. Furthermore, using an intelligent semantic sorting method based on pre-trained language models, it outputs a list of candidate patents related to the target functional behaviors. The effectiveness of this approach is illustrated in Tables 1. This integrated system not only enhances the accuracy of retrieving relevant patents but also significantly streamlines the process of identifying patents that are most pertinent to the users' specified functional requirements.

Table 1. Collection of Semantic Search Patents.

Technological Objectives	Functional Behaviors	New patents (domain1)	New patents (domain1)
Objective 1	Functional Behaviors 1	Patent 1 Patent 6	Patent 3 Patent 5
Objective 2	Functional Behaviors 2	Patent 4 Patent 9 Patent 11	Patent 5 Patent 7
Objective 3	Functional Behaviors 3	Patent 12 Patent 13	Patent 14 Patent 15 Patent 16

2.2 Identifying Technology Opportunities

Quantitative Analysis of Technological Opportunities

The Spherical Fuzzy Sets (SFS) theory provides a novel framework for addressing issues of ambiguity and uncertainty. This theory expands traditional fuzzy sets by introducing spherical membership degrees, thereby enabling a more precise representation of membership uncertainty for elements [16]. Suppose U is a finite, non-empty universal set. A spherical fuzzy set A on the universal set U is described by a tuple $(x, \mu_s(x), \nu_s(x), \pi_s(x))$, where x is an element of U, and $\mu_s(x), \nu_s(x)$, and $\pi_s(x)$ respectively represent the membership, neutrality, and non-membership degrees of x to A. These three quantities are calculated by the functions μ_s, ν_s, and π_s defined on U. The relationships among these measures are constrained by the condition that for any x in U, it satisfies $0 \leq \mu_s(x)^2 + \nu_s(x)^2 + \pi_s(x)^2 \leq 1$. This condition ensures a balance and consistency among the membership, neutrality, and non-membership degrees.

The spherical fuzzy number within a non-empty set can be represented as Formula 4.

$$P = \{x, \mu_s(x), \nu_s(x), \pi_s(x) | x \in X\} \tag{4}$$

For spherical fuzzy numbers, the Spherical Weighted Arithmetic Mean (SWAM) and the Spherical Weighted Geometric Mean (SWGM) are introduced. By incorporating the aforementioned set of weights, we derive the computational formulas for SWAM and SWGM, as presented in Eq. 5:

$$\begin{cases} SWAM(\omega_1, \omega_2, ..., \omega_n) = \left(\mu_{sp}, \nu_{sp}, \pi_{sp}\right) \\ SWGM(\omega_1, \omega_2, ..., \omega_n) = \left(\mu_{sq}, \nu_{sq}, \pi_{sq}\right) \end{cases} \tag{5}$$

Following the introduction of the SWAM (Spherical Weighted Arithmetic Mean) and SWGM (Spherical Weighted Geometric Mean) operators, the membership degree μ_{sp}, non-membership degree ν_{sp}, and hesitation degree π_{sp} of the SWAM operator, along with the membership degree μ_{sq}, non-membership degree ν_{sq}, and hesitation degree π_{sq} of the SWGM operator, all satisfy Eq. 6.

$$\begin{cases} \mu_{sp} = \sqrt{1 - \prod_{i=1}^{n} \left(1 - \nu_{si}^2\right)^{\omega_i}} \\ \nu_{sp} = \prod_{i=1}^{n} \nu_{si}^{\omega_i} \\ \pi_{sp} = \sqrt{\prod_{i=1}^{n} \left(1 - \mu_{si}^2\right)^{\omega_i} - \prod_{i=1}^{n} \left(1 - \mu_{si}^2 - \pi_{si}^2\right)^{\omega_i}} \\ \mu_{sq} = \prod_{i=1}^{n} \mu_{si}^{\omega_i} \\ \nu_{sq} = \sqrt{1 - \prod_{i=1}^{n} \left(1 - \mu_{si}^2\right)^{\omega_i}} \\ \pi_{sq} = \sqrt{\prod_{i=1}^{n} \left(1 - \nu_{si}^2\right)^{\omega_i} - \prod_{i=1}^{n} \left(1 - \nu_{si}^2 - \pi_{si}^2\right)^{\omega_i}} \end{cases} \tag{6}$$

Enhancing the MULTIMOORA Method for Ranking Technological Opportunities

Due to the superior stability and applicability of the MULTIMOORA method compared to other multi-attribute decision-making methods such as TOPSIS and VIKOR, the MULTIMOORA method is employed in conjunction with Spherical Fuzzy Set Theory to rank technological opportunities [17]. The specific steps for ranking technological opportunities using the combination of the MULTIMOORA method and Spherical Fuzzy Set Theory are as follows:

Calculation of Weight Factor Set

Construct the deviation fuzzy number matrix. Subsequently, calculate the weight factor set using the formula presented as Eq. 7:

$$\varpi_{kv} = \left| \frac{\sum_{u=1}^{3} (\tilde{\mu}_{skuv} - \tilde{v}_{skuv})^2 - \sum_{u=1}^{3} (\tilde{\pi}_{skuv} - \tilde{v}_{skuv})^2}{\sum_{u=1}^{3} \sum_{v=1}^{n} (\tilde{\mu}_{skuv} - \tilde{v}_{skuv})^2 - \sum_{u=1}^{3} \sum_{v=1}^{n} (\tilde{\pi}_{skuv} - \tilde{v}_{skuv})^2} \right| \tag{7}$$

ϖ_{kv} represents the weight factor set derived from the deviation of the v-th technological subsystem as assessed by the k-th expert.

Ranking Technological Opportunities

Using the ratio model for opportunity ranking:

$$\begin{cases} \tilde{\mu}_{skuv1} = \sqrt{\left|1 - \prod_{u=1}^{3} \prod_{v=1}^{n} \left|1 - \tilde{\mu}_{skuv}^2\right|^{\omega_i}\right|} \\ \tilde{v}_{skuv1} = \prod_{u=1}^{3} \prod_{v=1}^{n} \tilde{v}_{skuv}^{\omega_i} \\ \tilde{\pi}_{skuv1} = \sqrt{\left|\prod_{u=1}^{3} \prod_{v=1}^{n} \left|1 - \tilde{\mu}_{skuv}^2\right|^{\omega_i} - \prod_{u=1}^{3} \prod_{v=1}^{n} \left|1 - \tilde{\mu}_{skuv}^2 - \tilde{\pi}_{skuv}^2\right|^{\omega_i}\right|} \end{cases} \tag{8}$$

Based on Eq. 8, the spherical fuzzy number ζ_{km} corresponds to the opportunity assessment of the m-th sub-technology by the k-th expert. The ranking of opportunities is determined by the results of this comparison.

Using the reference point method for opportunity ranking:

Calculate the spherical distance based on the spherical fuzzy number Q_{kuv}, the corresponding weight factor set $w_k = (\varpi_{k1}, \varpi_{k2}, ..., \varpi_{kn})$, and the opportunity mode assessment vector $r_k = (r_{k1}, r_{k2}, ..., r_{kh})$ as illustrated in the Eq. 9:

$$d_{km} = r_{km} \max_{\substack{1 \le u \le 3 \\ 1 \le v \le n}} \frac{2\varpi_{kv}}{\pi} arccos\left\{1 - \frac{1}{2}\left[(\tilde{\mu}_{skuv} - 1)^2 + \tilde{v}_{skuv}^2 + (\tilde{\pi}_{skuv} - 1)^2\right]\right\} \tag{9}$$

Based on this calculation, the distance value d_{km} corresponding to the opportunity assessment of the m-th sub-technology by the k-th expert is determined, which further facilitates the ranking of technological opportunities.

Using the ratio model for opportunity ranking:

$$\begin{cases} \tilde{\mu}_{skuv2} = \prod_{u=1}^{3} \prod_{v=1}^{n} \tilde{\mu}_{skuv}^{\omega_i} \\ e\tilde{v}_{skuv2} = \sqrt{\left|1 - \prod_{u=1}^{3} \prod_{v=1}^{n} \left|1 - \tilde{v}_{skuv}^2\right|^{\omega_i}\right|} \\ \tilde{\pi}_{skuv2} = \sqrt{\left|\prod_{u=1}^{3} \prod_{v=1}^{n} \left|1 - \tilde{v}_{skuv}^2\right|^{\omega_i} - \prod_{u=1}^{3} \prod_{v=1}^{n} \left|1 - \tilde{v}_{skuv}^2 - \tilde{\pi}_{skuv}^2\right|^{\omega_i}\right|} \end{cases} \tag{10}$$

According to the Eq. 10, the spherical fuzzy number θ_{km} represents the opportunity assessment for the m-th sub-technology by the k-th expert. The ranking of technological opportunities is then determined based on the results of this comparison.

3 Case Study

3.1 Application Background and Needs Analysis

With the widespread application of polyethylene (PE) piping systems in water supply, gas distribution, oil transportation, and coastal aquaculture, the workload for pipeline installation has significantly increased. Traditionally, the heating plate and cutting tool of welding machines are positioned separately. After the pipe is cut, the cutting tool is removed and replaced with the heating plate to complete the butt welding process. Both the cutting tool and the heating plate are electrically operated, and in confined spaces, the alternating replacement of these components poses potential safety risks to operators.

In the heating plates of existing plastic pipe fusion welding machines, the heating tube is arranged circumferentially, with temperature measurement conducted at a single point to serve as the input for temperature control. As a result, the heating of the entire tube is regulated based solely on the temperature at that point, making it difficult to maintain uniform temperature across the heating plate, which adversely affects the welding quality. Furthermore, as shown in Fig. 3, these heating plates typically use fixed-shaped heating elements, which cannot be adjusted according to specific needs. This limitation is particularly pronounced when welding elliptical cut surfaces formed by pipes of various shapes. Moreover, since different pipe materials require distinct heating methods and heat distribution, the existing heating plates fail to meet these demands..

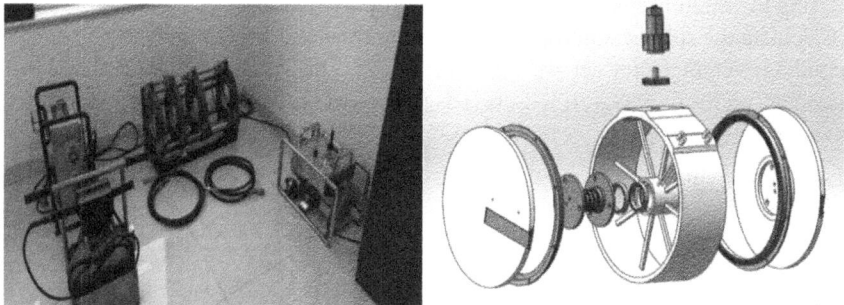

Fig. 3. Hot melt welder (left) and Integrated milling heating plate (right).

3.2 Transforming Technology into Functional Requirements

We first explored various data sources by searching for documents containing the following keywords: "fusion welding machine," "pipe fusion welding," and "fusion welding machine design." We then selected a total of 40 samples from each type of source, including encyclopedic entries, scientific articles, and web articles. These samples were used to train the "models--bert-base-chinese" model. The trained model was subsequently employed to perform similarity detection between the abstracts and the technical details of patents. The abstracts represent abstract expressions, while the technical details provide specific technical information. The trained model demonstrated higher similarity scores and a more compact score distribution, indicating its improved effectiveness in understanding and encoding domain-specific texts. As shown in Fig. 4, the majority of the patent scores clustered around 0.99, reflecting a high degree of consistency. Further confirmation of this improvement was obtained through t-test and p-value analyses, which showed that the enhancement was not coincidental but a direct result of the model's training and optimization. t-Statistic: -13.5228, p-value: 8.93411.

Fig. 4. Training Results

The technological objectives aim to enhance the consistency and reliability of welding, improve production efficiency, provide uniform heating with adjustable heating plates based on actual demands, enable multi-angle and multi-position adjustments of the cutting tool, ensure cutting precision and repeatability, improve response speed and operational efficiency, achieve precise angle control with adaptive welding functions, and replace hydraulic and constant temperature pumping systems with mechanical alternatives. Following the transformation of these objectives into functional requirements:

Functional Behavior 1: The control system should support multi-angle and multi-position adjustments to accommodate complex welding demands, thereby enhancing the consistency and reliability of welding while ensuring cutting precision and repeatability.

Functional Behavior 2: The heating system of the hot melt welding machine needs to be capable of adjusting the shape and distribution of the heating plates according to the different shapes and material requirements of the workpieces, to achieve uniform heating.

Functional Behavior 3: Design a folding angle control system driven mechanically, replacing traditional hydraulic systems. This system should include high-precision servo motors and mechanical linkages to precisely control the angle and speed of the folding operation in unfinished welded tubes.

These are designated as Y1, Y2, and Y3, respectively.

3.3 Semantic Retrieval of Patents for Functional Behaviors

Using the Patsnap semantic search tool, patents semantically related to specific functional behaviors are identified and ranked in descending order of semantic relevance. As shown in Table 2. In this case, the semantic relevance scores for the functional behaviors are as follows: Y1 has the lowest semantic relevance at 0.91, Y2 at 0.93, and Y3 at 0.90. This method ensures that the patents retrieved are closely aligned with the technical specifications and functional requirements outlined in the project.

Table 2. Overview of Semantic Search Results.

Functional Behavior	Example
Y1	CN109159007B
	CN206840193U
	CN206870150U
	CN206912337U
	CN215588615U
	CN220296609U
	CN108555776A
	WO2023168848A1
	CN211589448U
	CN204085916U
Y2	CN202318944U
	CN209649505U
	CN203236711U
	CN2712618Y
	CN114834056A
	CN208290506U
	CN214082919U
	CN217435061U
	CN114030188A
	CN219789322U
Y3	CN215034743U
	CN218256735U
	CN116944371A
	GB848464A
	CN101370646B
	CN112815177A
	CN102866665B
	CN113828891A
	CN102866665A

Ranking Technological Opportunities and Feasibility Verification

In this context, Y_{kw} denotes the w-th patent associated with the k-th requirement. The ranking of technological opportunities using the MULTIMOORA method combined with Spherical Fuzzy Set theory is detailed in the Table 3 provided.

Table 3. TO Scores for Y1 and Y3

	Proportional System Score	Standard Deviation Score	Full Multiplication Model Score	Composite Score
Y16	5	2.45	8	10.51
Y15	5	2.45	3	5.51
Y17	3	1.41	3	5.51
Y110	3	1.41	3	5.51
Y11	3	1.41	1	2.6
Y14	3	1.41	1	2.6
Y19	2	1	1	2.6
Y12	1	0	1	1
Y13	1	0	1	1
Y18	1	0	1	1
Y38	3	0.49	8	10.54
Y35	2	0.45	4	5.54
Y36	2	0.45	4	5.54
Y37	2	0.45	4	5.54
Y39	2	0.45	4	5.54
Y33	1	0.35	2	2.65
Y31	0	0	1	1
Y32	0	0	1	1
Y34	0	0	1	1

Because the heating system needs to adjust the shape and distribution of the heating plates according to the shapes of different PE pipes, and the support structure of the heating plates is controlled and adjusted by the control system, the identification of technological opportunities for Y3 should build upon the foundation set by Y1. Specific data is provided in the Table 4 below:

Table 4. TO Scores for Y2 Building upon Y1

Heading level	Y21	Y22	Y23	Y24	Y25	Y26	Y27	Y28	Y29	Y210
Y11	1.0; 0.3; 2.0; 2.7	0.0; 0.0; 1.0; 1.0	0.0; 0.0; 1.0; 10.1	2.0; 0.4; 4.0; 5.6	2.0; 0.4; 4.0; 5.61	3.0; 0.458; 8.0; 10.54	2.0; 0.40; 4.0; 5.62	0.0; 0.0; 1.0; 10.2	2.0; 0.40; 4.0; 5.63	2.0; 0.40; 4.0; 5.64
Y12	3; 0.458; 8; 10.5	2; 0.4; 4; 5.6	2; 0.4; 4; 5.6	4; 0.49; 16; 19.5	4; 0.49; 16; 19.5	5; 0.5; 32; 36.5	4; 0.49; 16; 19.5	2; 0.4; 4; 5.6	4; 0.49; 16; 19.5	4; 0.49; 16; 19.5
Y13	4; 0.49; 16; 19.5	3; 0.458; 8; 10.5	3; 0.458; 8; 10.5	5; 0.5; 32; 36.5	5; 0.5; 32; 36.5	5; 0.5; 32; 36.5	5; 0.5; 32; 36.5	3; 0.458; 8; 10.5	5; 0.5; 32; 36.5	5; 0.5; 32; 36.5
Y14	3; 0.458; 8; 10.5	2; 0.4; 4; 5.6	2; 0.4; 4; 5.6	4; 0.49; 16; 19.5	4; 0.49; 16; 19.5	5; 0.5; 32; 36.5	4; 0.49; 16; 19.5	2; 0.4; 4; 5.6	4; 0.49; 16; 19.5	4; 0.49; 16; 19.5
Y15	3; 0.458; 8; 10.5	2; 0.4; 4; 5.6	2; 0.4; 4; 5.6	4; 0.49; 16; 19.5	4; 0.49; 16; 19.5	5; 0.5; 32; 36.5	4; 0.49; 16; 19.5	2; 0.4; 4; 5.6	4; 0.49; 16; 19.5	4; 0.49; 16; 19.5
Y16	1; 0.3; 2; 2.7	0; 0; 1; 1	0; 0; 1; 1	2; 0.4; 4; 5.6	2; 0.4; 4; 5.6	3; 0.458; 8; 10.5	2; 0.4; 4; 5.6	0; 0; 1; 1	2; 0.4; 4; 5.6	2; 0.4; 4; 5.6
Y17	3; 0.458; 8; 10.5	2; 0.4; 4; 5.6	2; 0.4; 4; 5.6	4; 0.49; 16; 19.5	4; 0.49; 16; 19.5	5; 0.5; 32; 36.5	4; 0.49; 16; 19.5	2; 0.4; 4; 5.6	4; 0.49; 16; 19.5	4; 0.49; 16; 19.5
Y18	1; 0.3; 2; 2.7	0; 0; 1; 1	0; 0; 1; 1	2; 0.4; 4;5.6	2; 0.4; 4;5.6	3; 0.458; 8; 10.5	2; 0.4; 4; 5.6	0; 0; 1; 1	2; 0.4; 4; 5.6	2; 0.4; 4;5.6
Y19	2; 0.4; 4; 5.6	1; 0.3; 2; 2.7	1; 0.3; 2; 2.7	3; 0.458; 8; 10.5	3; 0.458; 8; 10.5	4; 0.49; 16; 19.5	3; 0.458; 8; 10.5	1; 0.3; 2; 2.7	3; 0.458; 8; 10.5	3; 0.458; 8; 10.5
Y110	3; 0.458; 8; 10.5	2; 0.4; 4; 5.6	2; 0.4; 4; 5.6	3; 0.458; 8; 10.5	4; 0.49; 16; 19.5	5; 0.5; 32; 36.5	4; 0.49; 16; 19.5	2; 0.4; 4; 5.6	3; 0.458; 8; 10.5	4; 0.49; 16; 19.5

3.4   Application of Technological Opportunities

Alternative Solutions and Their Optimization

Some technological opportunities, due to their cross-disciplinary applications or specific requirements in different contexts, cannot be directly implemented. To ensure the practicality and effectiveness of these technologies, they must be optimized according to specific circumstances. The details are shown in the Table 5 below.

Table 5. List of Improved Partial Alternative Solutions.

No.	Issue to be Improved	Optimization Tool	Optimized Solution
1	Overlap exists between the workspace of the heating plate and milling device, though not in their operating times	Trimming	Trim the supporting arm of one, combining both devices into a single unit
2	The heating plate operates in different regions for varying pipe angles	Principle of Time Separation	Implement a dot matrix heating system to cater to different heating zones
3	Increased functionality is needed within a limited space	Law of Increasing Ideality	Enable composite functions in some components to enhance functionality
4	Different components need to avoid motion interference	Law of Uneven Development of Subsystems	Ensure independent movement among components to prevent interference
5	Adjustment of cutting speed is necessary for different materials and cutting requirements	Evolutionary Pathway of Increasing Coordination	Install a speed regulator in the cutting device's motor to adjust the cutting speeds
6	The cutting tool operates in different areas for varying pipe tasks	Inventive Principle 2: Separation Inventive Principle 13: Inversion	Separate the cutting tool, creating an independent workspace and control system
7	The design of circular bending adds complexity to the heating plate design	Inventive Principle 16: Partial or Excessive Actions	Adopt a folding bending technique to simplify the design

Final Design

Drawing on the highest scoring CN2712618Y, the heating plate utilizes a disc structure equipped with a dense matrix of resistive metal pieces. This configuration allows for precise calculations of the ellipse shape based on wall thickness and bevel angle, ensuring accurate heating to the specified temperature. This design emulates the spatial filling properties of high-temperature liquids found in traditional hot melt welding machines, such as CN114834056A, ensuring uniform heating and quality of the weld joints. The tool design is inspired by the highest scoring CN220296609U under category Y1. However, to avoid increasing the complexity of the support arm and folding mechanism, invention principles 13 and 2 are utilized. The tool moves independently to complete the circumferential cutting motion. The support arm, inspired by the second highest scoring CN215588615U under Y1, is designed with three degrees of rotational freedom, one of which is for the bevel angle. These three degrees work together to perform precise positional control. After one end is beveled, the system retracts to perform a 180-degree rotation around the axis, positioning the other end at the set folding angle, which is twice the bevel angle. As shown in Fig. 5.

In the bending functionality design, we adopt a folding bending approach. If a circular arc bending design were used, the complexity of the heating plate design would significantly increase. Drawing from the third highest scoring CN112815177A and using invention principle 16, we employ a simplified folding bending approach. Although this bending method does not achieve an ideal circular arc shape, it is more practical and feasible in design, ensuring the functionality and reliability of the entire system.

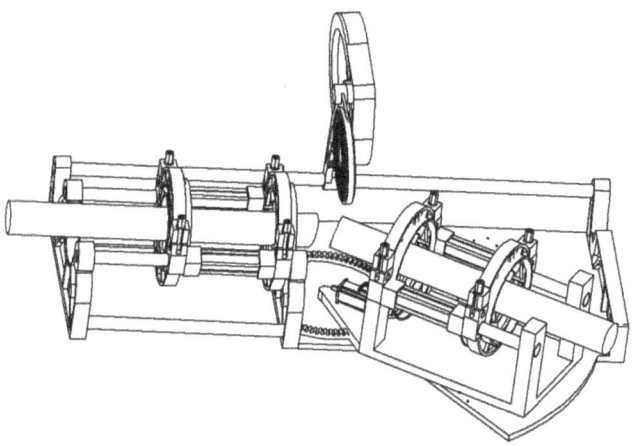

Fig. 5. Final solution diagram of the hot melt welding machine.

4   Conclusions

Product innovation is a key factor in enhancing a company's competitiveness. The identification of technological opportunities provides strategic direction for the company. Using the BERT model, we transformed multiple technological objectives into functional behaviors and identified patents that meet these functions through semantic relevance search. Spherical fuzzy sets helped us transition from qualitative to quantitative analysis. Ultimately, using the MULTIMOORA method, we identified single or multiple interacting technological opportunities.

Product innovation is a key factor in enhancing a company's competitiveness, and the identification of TO provides strategic direction for firms. Scholars have been working to make the TOA method more objective and effective by reducing the reliance on expert involvement and improving the interpretation and evaluation of opportunities throughout the process. By utilizing the BERT model to process text, multiple technological objectives were translated into functional behaviors, and patents satisfying these functional behaviors were identified through semantic relevance retrieval. Spherical fuzzy sets were employed to transform qualitative analysis into quantitative analysis. Finally, the MULTIMOORA method was used to identify single or multiple interacting technological opportunities.

First, the method proposed in this paper can handle domain-specific knowledge with greater precision, without the need to fine-tune the BERT model as in previous research. The MLM uses self-supervised learning, which does not require explicitly labeled data, whereas fine-tuning relies on supervised learning that necessitates task-specific labeled data. This approach simplifies the sample processing workflow. Second, the proposed method can identify potential technological opportunities from other fields. It is not based on domains or IPC classification, nor on clustering, but instead on semantic retrieval of functions. This approach eliminates the limitations of experts who are typically proficient in only one field and struggle to venture into others, significantly reducing the blind spots in innovation. Third, the method demonstrates the capability to handle interacting technological opportunities. Unlike traditional identification methods that are primarily limited to finding market gaps or individual independent technological opportunities, this method considers the complexity and interactivity of technologies. In practical applications, technological opportunities rarely exist in isolation; they are often interconnected and mutually influential. Therefore, this method can more comprehensively capture and analyze these interacting technological opportunities, providing stronger support for technological innovation and strategic planning.

Despite the numerous contributions of the proposed method, certain limitations remain. First, as TOA is a crucial part of business activities, it must fully account for the influence of competitors and collaborators. In the future, the supplementation and improvement of relevant data sources will be necessary. Second, the BERT model is still considered a black box in its application, with the relationships between internal technologies remaining unclear. While machine learning tools can provide results, the specific operational processes are difficult to interpret. Third, when selecting new technologies, it is essential not only to consider the factors related to competitors and collaborators but also to focus on evaluating the company's own technological capabilities and comprehensively assessing the risks and benefits associated with the technology to determine the most suitable technological opportunities. Although the proposed method offers valuable assistance, these issues warrant further in-depth research.

References

1. Coccia, M.: Sources of technological innovation: radical and incremental innovation problem-driven to support competitive advantage of firms. Technol. Anal. Strateg. Manag. **29**(9), 1048–1061 (2017)
2. Liu, L., Jiang, Z.: Influence of technological innovation capabilities on product competitiveness. Ind. Manag. Data Syst. **116**(5), 883–902 (2016)
3. Ramadani, V., et al.: Product innovation and firm performance in transition economies: a multi-stage estimation approach. Technol. Forecast. Soc. Change **140**, 271–280 (2019)

4. Possas, M.L., Salles-Filho, S., da Silveira, J.M.: An evolutionary approach to technological innovation in agriculture: some preliminary remarks. Res. Policy **25**(6), 933–945 (1996)

5. Wang, J., et al.: Development of technology opportunity analysis based on technology landscape by extending technology elements with BERT and TRIZ. Technol. Forecast. Soc. Change **191**, 122481 (2023)

6. Han, X., et al.: Technology opportunity analysis: combining SAO networks and link prediction. IEEE Trans. Eng. Manag. **68**(5), 1288–1298 (2019)

7. Rodriguez, A., et al.: Patent clustering and outlier ranking methodologies for attributed patent citation networks for technology opportunity discovery. IEEE Trans. Eng. Manag. **63**(4), 426–437 (2016)

8. Teng, F., et al.: Technology opportunity discovery of proton exchange membrane fuel cells based on generative topographic mapping. Technol. Forecast. Soc. Change **169**, 120859 (2021)

9. Cozzens, S., et al.: Emerging technologies: quantitative identification and measurement. Technol. Anal. Strateg. Manag. **22**(3), 361–376 (2010)

10. Cho, Y., et al.: Identifying technology opportunities for electric motors of railway vehicles with patent analysis. Sustainability **13**, 2424 (2021)

11. Yun, S., et al.: Technological trend mining: identifying new technology opportunities using patent semantic analysis. Inf. Process. Manag. **59**(4), 102993 (2022)

12. Pershina, R., Soppe, B., Thune, T.M.: Bridging analog and digital expertise: cross-domain collaboration and boundary-spanning tools in the creation of digital innovation. Res. Policy **48**(9), 103819 (2019)

13. Kim, J., Kim, S., Lee, C.: Anticipating technological convergence: link prediction using Wikipedia hyperlinks. Technovation **79**, 25–34 (2019)

14. Kumar, A., et al.: Link prediction techniques, applications, and performance: a survey. Phys. A: Stat. Mech. Appl. **553**, 124289 (2020)

15. Lou, W., Meng, J.: The diversity of canonical and ubiquitous progress in computer vision: a dynamic topic modeling approach. Inf. Process. Manag. **60**(3), 103238 (2023)

16. Büyüközkan, G., Karabulut, Y., Göçer, F.: Spherical fuzzy sets based integrated DEMATEL, ANP, VIKOR approach and its application for renewable energy selection in Turkey. Appl. Soft Comput. **158**, 111465 (2024)

17. Zhang, C., et al.: Intuitionistic fuzzy MULTIMOORA approach for multi-criteria assessment of the energy storage technologies. Appl. Soft Comput.Comput. **79**, 410–423 (2019)

AI-Aided Resource Mining Method for Idealization-Driven Product Innovation

Ranye Du[1,2]([✉]), Jianguang Sun[1,2], Runze Miao[1,2], and Delong Zhang[1,2]

[1] School of Mechanical Engineering, Hebei University of Technology, Hebei University of Technology, Tianjin 300401, China
18222724198@163.com

[2] National Engineering Research Center for Technology Innovation Method and Tool, Hebei University of Technology, Hebei University of Technology, Tianjin 300401, China

Abstract. The availability of resources in design plays a crucial role in innovative design. The importance of available resources increases as the solution to the problem approaches the Ideal Final Result (IFR). This paper introduces a method for mining resources in the product innovation process by integrating the Theory of Inventive Problem Solving (TRIZ) and AI, and demonstrates TRIZ's application in various fields.

Firstly, through AI tools to obtain research hotspots to formulate product design goals, through the training of ChatGPT, with the help of AI tools to achieve the mining of internal resources. Additionally, an AI algorithm is employed to analyze patents and identify external available resources. We evaluate the value of these mined resources using a Technique for Order Preference by Similarity to Ideal Solution (TOPSIS) comprehensive evaluation model that combines analytic hierarchy process with entropy weight method to select optimal available resources. Multi-criteria evaluation methods and AI technology are applied to resource mining and selection for product innovation.

Keywords: AI and Triz · Topsis · AI-assisted Resource Mining

1 Introduction

Since the 1990s, the goal of enterprise production has shifted from an obsessive pursuit of production efficiency to innovation in product functionality. With the increasing scarcity of global energy and the depletion of resources, the rational use of resources and "green" innovation have become crucial. Behind this shift, every enterprise faces the challenge of how to improve production efficiency and how to better design products. For a more extended forecast, prediction becomes inherently challenging due to the imminent paradigm shift we are currently witnessing in the evolution of TRIZ. The advent of the artificial intelligence (AI) era marks a significant turning point for TRIZ from a scientific standpoint [1].

Jochen Wessne [2] describes an example that could be mapped to the third level of increasing ideality, where the rational use of resources can achieve sustainable development. The invention, production, and use of a product require the involvement of many

© IFIP International Federation for Information Processing 2025
Published by Springer Nature Switzerland AG 2025
D. Cavallucci et al. (Eds.): TFC 2024, IFIP AICT 735, pp. 147–164, 2025.
https://doi.org/10.1007/978-3-031-75919-2_9

resources, and the resources available in the design play an important role in product innovation.

Currently, there is limited research on resource mining methods, Through the study and research of relevant literature, it has been found that existing resource mining methods mainly involve establishing a system black-box model or a system functional model to identify the functional-result-effect-resource interrelationships within the system. This helps designers locate resources. For example, Zuguang Li [3] proposed different resource mining methods based on various product innovation types by performing functional analysis on the system. Yue Hu [4] directly mined usable resources through causal analysis and expanded these resources by performing functional decomposition and constructing effect chains to identify the system's usable resources. Nan Ma [5] completed the extraction of available resources within a system by constructing a system functional structure and functional model. Additionally, by studying patent texts to obtain knowledge resources, the Apriori algorithm was employed to extract and identify innovative knowledge resources within the patents. Hui Dong [6] proposed a resource analysis process for the conceptual design of mechanical products based on function-effect-resource-component. By integrating usable resources through the interrelationships between components, innovative solutions can be obtained.

These methods largely depend on the analyst's individual expertise and mostly involve identifying resources and then finding usable resources to improve the product. Consequently, many resources identified in previous analyses are ineffective. Therefore, can we, in the context of knowing the product design requirements in advance, conduct product innovation to mine resources, improve the efficiency of resource mining, and find product innovation solutions that can be applied to various different scenarios? Integrating AI into the resource mining process and using multi-criteria evaluation methods to plan the application of resources in different scenarios could potentially address this issue.

Based on identifying product design hotspots through AI, this paper develops an AI-assisted resource mining method. The method primarily involves training ChatGPT to discover internal resources of the product and using ChatGPT to construct search queries for building a patent database. Patents contain numerous innovative solutions, and these innovative principles are information resources within the invention resources. By exploring patent texts, the corresponding innovative principles can be obtained. These queries are used to retrieve principles applicable to product innovation, and the retrieved patents are subjected to cluster analysis to identify principle terms and corresponding invention resources that can be used for product innovation. The method completes the extraction of external resources for the product. By employing the AHP-EWM-TOPSIS combined weighting method, the comparison values of different innovation schemes are calculated. These comparison values are used to determine the resources utilized by the product under different scenarios. This approach also serves as an evaluation method to determine product innovation schemes and application scenarios based on resources.

2   Related Research

2.1   TRIZ: Resources

TRIZ theory defines the resources that can be used for innovation as inventive resources and evolutionary resources [7]. Inventive resources refer to the materials available in the system and the environment; they can also be described as the totality of all materials, energy, and information that can be utilized by human knowledge and application [8]. Inventive resources can also be divided into internal resources and external resources. Internal resources refer to the resources that exist within the time and area where the conflict occurs. External resources are the resources that exist outside the time and area of the conflict [7].

Iouri Belski [9] points out that the basis of innovation is to use resources to satisfy needs, and a clear understanding of needs and resources is essential for the survival of the enterprise. The focus of industrial design is innovation, which is a process of discovering and solving problems. Utilizing TRIZ to solve problems and using resources to refine and improve product design [10]. The purpose of this paper is to explore the inventive resources that can be used in the process of product innovation, such as improving the material and information available for the product.

2.2   Ma

Morphological analysis (MA) is used to address multi-dimensional, usually non-quantifiable, problem complexes [11]. When constructing a morphological matrix, a complex problem can be divided into multiple dimensions., each dimension can then be divided into several forms to depict the characteristics of the entire system [12]. Identify related factors from multi-dimensional events and arrange them into a matrix diagram. Then, analyze the problem based on the matrix diagram, simplifying and clarifying the complex problem. In the context of innovative product design, MA can be combined with TRIZ to explore more potential systematic solutions.

2.3   TRIZ and AI

With the continuous development of artificial intelligence technology, AI has permeated various aspects of human daily life. TRIZ, the Theory of Inventive Problem Solving, aims to address a range of design issues. As AI technology advances, the combination of TRIZ and AI has become a possibility. Stelian Brad [13] explored the flexibility and durability of TRIZ's development, introducing the future trend where TRIZ will enter an era of integration with computers, driven by AI innovation. Nicolas Douard [14] introduced a method within the TRIZ framework that combines natural language processing models in AI with TRIZ to solve multidisciplinary problems. Simon Dewulf [15] presented a method of integrating AI into text mining to help designers search for patent knowledge, better aligning with TRIZ's inventive principles. This paper will explore the application of ChatGPT in assisting designers in product design and uncovering untapped resources during the process of innovative product design.

3 Proposed Method

3.1 ChatGPT and TRIZ

For complex products, identifying resources usually requires manual analysis and thought, which often consumes significant manpower. With the continuous development of artificial intelligence networks, using AI to assist in this work has become a possibility. ChatGPT can provide logically clear solutions and plans based on the user's needs and objectives. The question now is whether it is possible to combine TRIZ with AI in a reasonable way to enhance this process. After determining the product design objectives, product innovation requires the support of resources. The use of TRIZ tools and AI methods, combined with the mining of product resources. ChatGPT is a Generative Pre-trained Transformer, which is a deep neural network pre-trained with massive data. Through repeated learning, the model can master general natural language processing abilities.

3.2 Internal Resource Mining

Internal resources are those within the current technology system. The use of AI tools to find internal resources can be achieved by training ChatGPT to analyze the product (for ease of understanding, the use of an electrofusion welding machine, for example). However, the direct use of ChatGPT to find resources results in lower accuracy.

Therefore, this article will focus on how to train ChatGPT to explore the internal resources contained within a product. This can be achieved using ChatGPT's own Explores GPTs feature. When using ChatGPT alone, it is difficult to obtain clear and accurate internal resources of a product. For example, when I want to search for the internal spatial resources of an electrofusion welding machine, I directly ask ChatGPT: "I need you to identify the internal resources of the electrofusion welding machine, please help me list the spatial resources used by the electrofusion welding machine." The result was: internal space, heat dissipation ducts, cable routing space, control panel space, and capacitor module accommodation space. Upon analysis, this result does not rigorously represent the internal resources. Therefore, it is necessary to guide and train ChatGPT so that it can automatically provide the corresponding resources:

Creation: Use the functionality of constructing ChatGPT to make it the desired "Product Innovation Resource Explorer." This function allows you to define the features Chat-GPT needs to achieve based on your specific goals and guide ChatGPT to operate within a certain boundary to achieve these functions.

Training: Train ChatGPT and input the relevant definitions of resources from TRIZ so that ChatGPT can think within the specified definitions. For example, in the dialogue box, use: "In TRIZ, the definition of spatial resources is: position, order, and the space occupied by the system itself. I need you to find the spatial resources of the electrofusion welding machine based on the TRIZ definition of spatial resources." The result was: space inside the welding machine, space utilized by the welding sleeve, space for welding operations. Based on this, the accuracy of the results obtained by ChatGPT is further improved (see Fig. 1).

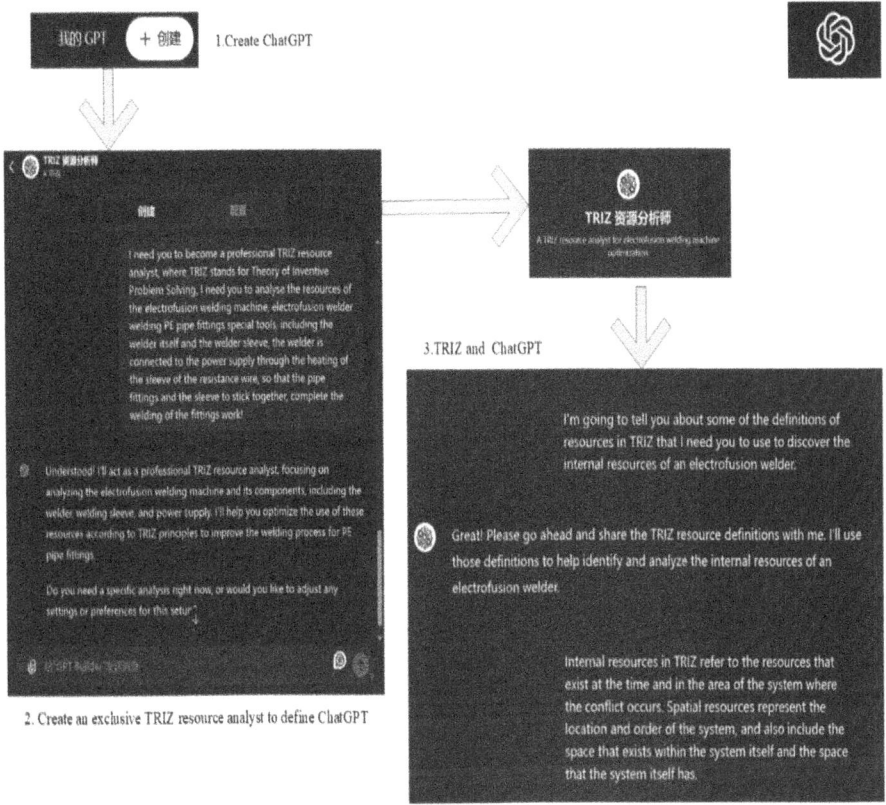

Fig. 1. ChatGPT Using Process

By guiding GPT, we can enhance its understanding of TRIZ-related knowledge, which it is relatively less familiar with. Hence, it is necessary to input more TRIZ-related resource knowledge to artificially train the GPT model to achieve better resource exploration results. The specific training process can be found in (see Fig. 2).

3.3 External Resource Mining

External resources are those that exist outside the time and region where the conflict occurs. Analogical design can apply knowledge from different regions to different areas, and generally speaking, the implementation of similar functions usually applies similar resources. This paper mines external resources based on analogical knowledge.

The system of a product contains numerous available resources. Identifying which resources have not yet been discovered and can be utilized to aid in product design is the first question that needs to be addressed. The focus of this paper is on how to provide problem-solving ideas for the design issues required by the product and, based on this, to identify usable resources to solve these issues. Patent databases are a concentrated point of design principles and cross-disciplinary design knowledge. By acquiring the

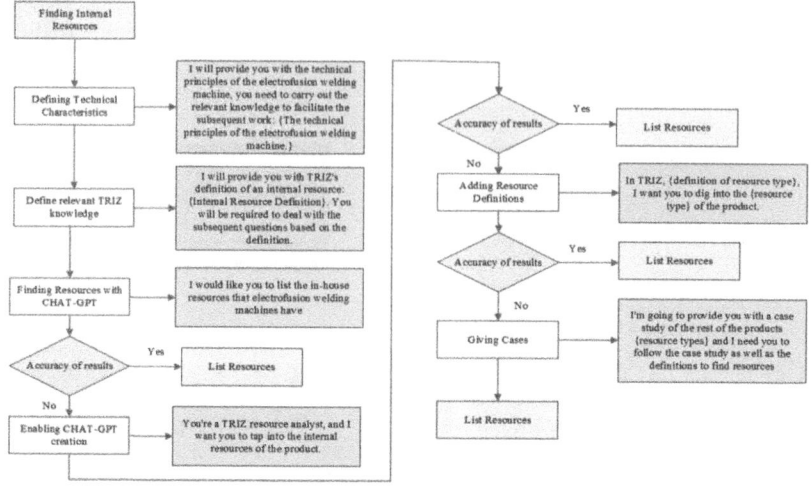

Fig. 2. ChatGPT Training Process

implicit design knowledge and information resources within patents, problems can be better solved, and better products can be developed (the patent database used in this paper is "Wisdom Patent", a patent search platform developed in China).

To obtain usable external resources and better utilize unknown information resources and untapped functional resources, this paper will use the innovative principles available in the patent database, employing the acquired information and functional resources to assist in product design. In the process of product design, other resources will be utilized to complete the design, thereby achieving the exploration of external resources. This paper will conduct text mining on the knowledge within the patent database to obtain information resources. Text mining is a computer science filed that incorporates various techniques such as data mining, natural language processing and knowledge management in order to uncover underlying information[16].

Preliminary Work: ChatGPT, as a natural semantic processing tool for AI, does not have the ability to directly search for patents but can be trained to obtain technical keywords for products and generate patent search terms to obtain relevant patents (see Fig. 3).

Patent Classification and Text Mining: The important objective of text mining is to find out previously unknown knowledge from a large content of texts [17].

TF-IDF Keyword Extraction: TF (Term Frequency) represents that the more frequently a word appears in a certain document, the more it can be a keyword represented by that document, and IDF (Inverse Document Frequency) indicates that these keywords can be used as a classification if only few texts contain these feature words. It can be described using the following formula:

$$W_j = [0.5 + 0.5(TF_j/TF_{max}) \times IDF \qquad (1)$$

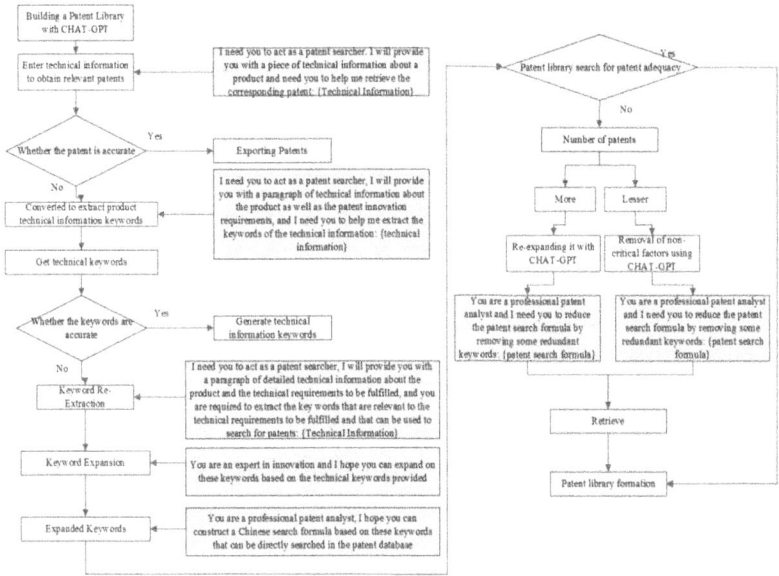

Fig. 3. Forming a Patent Library Training Process with ChatGPT

W_j represents the feature weights of j words in a text, TF_j represents the number of times word j appears in this text, TF_{max} represents the maximum number of times j appears in a single text in a collection of documents, and IDF represents the Inverse Document Frequency (IDF), which is of the form:

$$IDF = \log_N DF_j \qquad (2)$$

where N represents the total number of texts contained in a given collection DF_j represents the total number of texts in which j occurs in this collection of texts [10].

Morphological Matrix and Technology Roadmap: Morphological matrix is created for the keyword principles represented by different classifications. The technology roadmap is broadly used for planning technology, products, and markets because it provides action plans for achieving the goals (see Fig. 4). To calculate the similarity of patent keyword documents using cosine similarity, two vectors representing the documents calculate the angle between their vectors, and the cosine of the angle determines whether the two documents point in the same direction [18]:

$$cos\theta = \frac{A.B}{\|A\|\|B\|} = \frac{\sum_{i=1}^{n}(A_i \times B_i)}{\sqrt{\sum_{i=1}^{n}(A_i)^2} \times \sqrt{\sum_{i=1}^{n}(B_i)^2}} \qquad (3)$$

By following the above steps, it is possible to uncover hidden keywords in patent texts and use morphological matrices to construct and characterize innovation principle terms. Traditional patent classification mainly relies on IPC codes, but to obtain purposeful innovation principles, text mining is required to extract the innovative knowledge within patents. This not only broadens the designer's knowledge space but also improves the utilization of design knowledge and increases opportunities for effective and innovative solutions [18]. The use of patent knowledge represents the reutilization of external information resources, which can help guide the use of other resources and facilitate the exploration of external system resources.

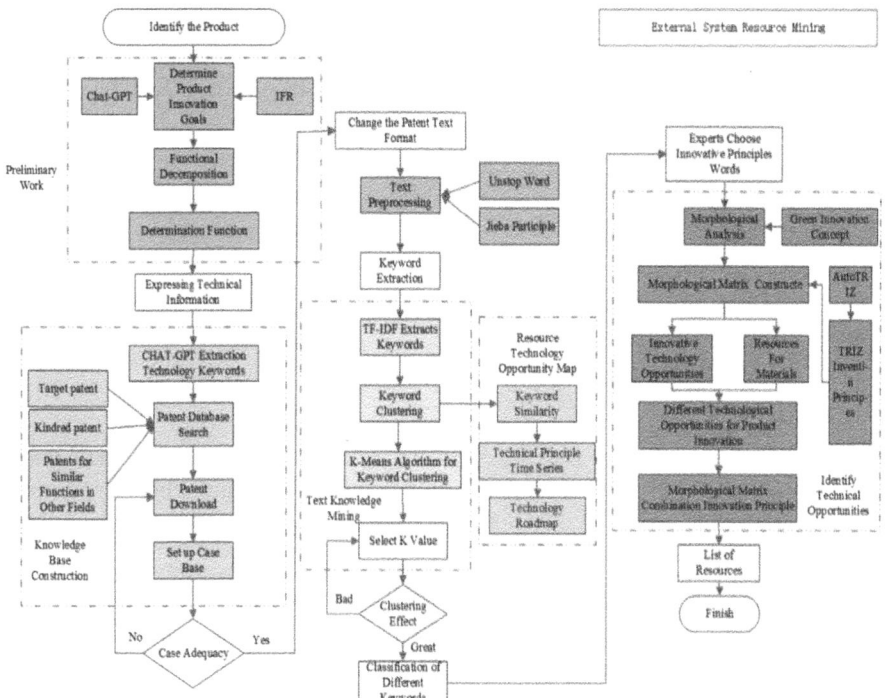

Fig. 4. External Resource Mining Process

3.4 Resource Evaluation

Innovative design can not only improve product functionality but also alter the use of different resources. Innovative design approaches are essential for achieving sustainability goals, as they can help reduce resource consumption, minimize waste and emissions, and promote ethical practices [19]. The proper use of available resources in design is important, and as the product innovation solution gets closer to the final ideal solution, the more important the available resources become.

The use of multi-criteria methods can assist in the selection of product innovation principles by considering various factors to choose appropriate innovation principles and reasonable resources. By establishing different evaluation criteria, algorithms can automatically determine which methods and types of resources to use in different scenarios.

This approach enhances product lifecycle sustainability and vitality, meeting long-term needs by adopting more sustainable and dynamic resources.

Depending on the specific design requirements of different products, the design system established in this paper provides a foundational solution to the research content, aiming to introduce a framework for designers to consider. While it is based on the study of the electrofusion welding machine, the resource evaluation system framework proposed in this paper can still serve as a conceptual guide for future designs. The product resource evaluation system process established in this paper is illustrated in the following steps:

Define the Problem and Target Area: The purpose of this step is to clearly identify the product's design problem and the target area that needs modification, providing clear objectives for the subsequent construction of evaluation criteria. For example, the design requirement could be for the electrofusion welding machine to be used in scenarios without an external power source.

Construction of the Evaluation Criteria System: The construction of the evaluation criteria system must adhere to principles of objectivity, comprehensiveness, practicality, and scientific rigor, considering the problem from multiple perspectives. For instance, in this paper, the selection of resource evaluation criteria references numerous studies and considers various factors such as the product's economic and environmental impact, to build the evaluation criteria system.

Weighting of Evaluation Criteria: Depending on different design requirements or the product's application in various scenarios, the selected resources will differ. For example, lightweight materials should be used for ease of transport in the wilderness, while environmental friendliness should be emphasized in scenarios requiring recycling. Therefore, the weight assigned to different criteria varies. While AHP (Analytic Hierarchy Process) is a common weighting method, it is highly subjective due to the need for experts to construct the decision matrix. To reduce the subjective influence of AHP, it can be combined with EWM (Entropy Weight Method), an objective weighting method.

Comprehensive Evaluation and Analysis: The TOPSIS (Technique for Order Preference by Similarity to Ideal Solution) method is used to establish the target evaluation system, which ranks options based on their proximity to the ideal solution. The constructed AHP-EWM-TOPSIS algorithm is used to comprehensively evaluate product resources, selecting the appropriate resources for different scenarios based on the constructed criteria and assigned weights.

Through the above methods, constructing a resource evaluation system can determine which innovation principles and resources should be used in different product scenarios, enabling a more objective selection of resources. The TOPSIS evaluation method is a comprehensive approach to evaluate various aspects of information. By using original data, it assesses different schemes, facilitating subsequent designers in selecting innovative solutions based on the evaluation results. In this article, a resource evaluation index system is established using the TOPSIS model. The final evaluation results are used to select product innovation solutions. Different index systems can be constructed, and weights can be set according to different scenarios to choose suitable innovation resources based on the context, generating various combinations of innovative solutions.

4 Case Study

4.1 Introduction and Setting Goals

Polyethylene (PE) pipelines are esteemed for their corrosion resistance and prolonged service life. A pivotal method in electric fusion welding involves encasing a heating sleeve wound with resistive wire around the pipe fittings to be welded. The electric fusion welding machine heats the sleeve, causing the resistive wire to melt the pipe fittings (see Fig. 5).

Fig. 5. Electrofusion welding machines and sleeves

ChatGPT, as an artificial intelligence that integrates a vast amount of knowledge, contains extensive databases and can, like a search engine, organize and combine knowledge from the web and various sources to present it. In this paper, ChatGPT is used to obtain innovative directions for electrofusion welding machines. The inquiry used is: "As a commonly used tool for welding PE pipes, I need you to help me identify the research hotspots and future innovation directions for electrofusion welding machines." The direction of technology development for using AI to acquire electrofusion welders is shown below Table 1.

Table 1. Electrofusion welding machine innovation goals

Development directions	Engineering objectives
Intelligent and automation	Utilizing advanced sensors and control systems to automatically adjust parameters to adapt to different welding conditions
Energy efficiency	Reducing energy consumption and minimizing environmental impact, thus decreasing costs
Portability	Expanding the range of applications for welding machines and implementing a multifunctional modular design for welding machines
Low pollution	Utilizing green and low-pollution materials to reduce environmental pollution

Integrated determination of innovative goals for welding machines: Utilizing welding machines in scenarios without a power supply (e.g., field, outdoor) while emphasizing cost reduction and the use of green, affordable materials are current focal points in electric fusion welding machine design.

4.2 Internal Resource Mining

In summary, following the internal resource extraction process outlined in this paper, ChatGPT can be trained to identify the internal resources of an electrofusion welding machine. As the methods for resource search have already been described in the previous sections, they will not be repeated here. The internal resources of the electrofusion welder are found by training the ChatGPT, and the internal resources are shown in the Table 2 below.

Table 2. Internal resources for electrofusion welding machine

Resource type	Resources
Information resources	Structural information such as welder size and weight; welder welding information; control system information
Field resources	Electric, thermal, mechanical, and magnetic fields
Functional resources	Preheating function, data analysis function, safety enhancement function, quality monitoring function
Temporal resources	Welding time, welding cycle time
Spatial resources	Space inside the welding machine, space utilist by the welding sleeve, space for welding operations
Material resources	Welding sleeve, structural components of the welding machine, Power modules for welding machines

4.3 External Resource Mining

ChatGPT Access to Technology Keywords: Using the search formula, we searched the corresponding patents in "Wisdom Sprout", downloaded a total of 2,517 relevant patents, formed a patent database, and obtained the relevant design principles in the database (I want you to act as a patent search expert. I will provide you with a piece of technical information, and you will need to extract keywords from it and expand those keywords to construct a Chinese patent search query that I can directly use in a patent database: An electrofusion welding machine, including the welding machine itself and a PE sleeve. I need an electrofusion welding machine that can be used outdoors without a power source.).

Text Processing and Keyword Extraction: Keyword word cloud map (see Fig. 6).

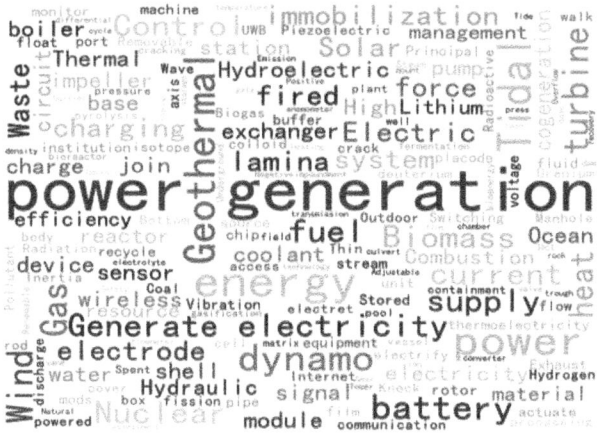

Fig. 6. Keyword word cloud map

Fig. 7. Keyword classification word cloud map

Nine categories are obtained after clustering various keywords and a word cloud is constructed for the keywords in the clusters as shown below (see Fig. 7).

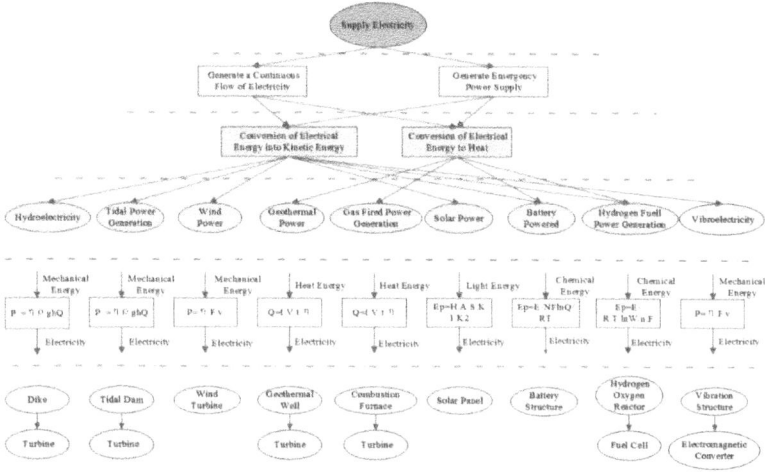

Fig. 8. Functional analysis of electrofusion welding machine

The electrofusion welding machine function diagram is as follows (see Fig. 8).

The morphological matrix based on the principles of electrofusion welding machine technology is shown in the Table 3 below.

Table 3. Morphological Matrix

Function	Value
Device power supply method	Tidal Turbines Ocean Energy Tidal Gates Tidal Energy Tidal Troughs Tidal Barriers Hydraulics Tidal Differential Generation Tidal Power Tidal Force Current Energy
	Geothermal Wells Geothermal Resources Steam Turbines Hot Water Pumps Geothermal Energy Thermal Fracking Underground Heat Exchangers Geothermal Fields Geothermal Heating Systems
	Lithium Ion Batteries Battery Packs Chargers Discharge Rates Cycle Life Deep Discharge Float Charge Charging Cycles Battery Recycling Sealed Batteries Open Batteries Nickel Hydride Batteries Lead Acid Batteries Lithium Ion Batteries Alkaline Batteries
	Green Energy Renewable Energy Self-Discharge Series Parallel Battery Stacks
	Wind turbines Blades Generators Towers Anemometers Wind vane Baseboards Converters Gearboxes
	Nuclear reactor Control rods Coolant Uranium fuel Nuclear fission Heavy water Reactor pressure vessel Nuclear power plants

(*continued*)

Table 3. (*continued*)

Function	Value
	Gas turbines Combustion chambers Natural gas Cogeneration Gas power stations High efficiency combustion Waste heat boilers
	Turbine Reservoir Generator Overflow channel Dam Pump storage plant
	Solar Panels Photovoltaic Power Generation Solar Thermal Solar Conversion Solar Power Plants Photovoltaic Inverters
	Solar modules Solar electric systems
	Adaptation: Rechargeable batteries Portable gas irrigation Hand-cranked generators Rechargeable batteries Mobile power supplies Taienergy chargers
Structure	Division: modular design of the welding machine, divided into power modules
Enclosure	Materials:Aluminium;Steel;Titanium;Polycarbonate;ABS;Polypropylene; Nylon(Polyamide); Glass Fibre Reinforced Plastics; Carbon fibre reinforced plastics

According to the constructed morphological matrix, there are $N = 11*10 = 110$. The experts select feasible solutions as indicated in the Table 4 below.

Table 4. Morphological Matrix

Principle	Lithiumion batteries; Solar generators; Nickel-hydrogen batteries; Nickel-cadmium batteries; Lead-acid batteries; Fuel cells; Outdoor generators
Material	Aluminium alloys; Stainless steel; Cast aluminium; Magnesium alloys; ABS; Copper Polycarbonate

Combined weighting of AHP and EWM as indicated in the Table 5 below. The weights in this scoring table are obtained through expert and ChatGPT evaluation.

Table 5. AHP and EWM

Indicators	AHP	EWM	Combined weighting
Multifunctionality of target resources	6.931	39.415	23.173
Volume of the product after using the target resource	23.165	7.998	15.5815
Power of the product after using the target resource	11.597	12.014	11.8055

(*continued*)

Table 5. (*continued*)

Indicators	AHP	EWM	Combined weighting
Recyclability of target resources	10.525	11.567	11.046
Cost of target resources	10.626	5.055	7.8405
Weight of the product after using the target resource	9.822	4.831	7.3265
Maintenance cost after using the target resource	8.087	5.87	6.9785
Pollution caused by the use of the target resource	6.883	3.974	5.4285
Multifunctionality of targeted resources	6.931	39.415	23.173

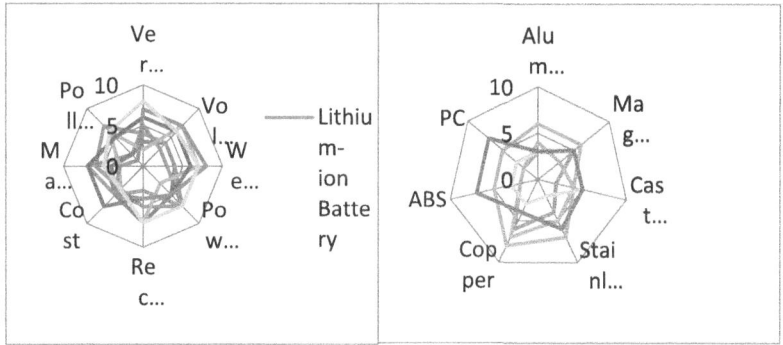

Fig. 9. Results of the evaluation of resources and innovation principles

The TOPSIS algorithm is used to calculate the resources used in the program to obtain the best resources to innovate the product (see Fig. 10) (Fig. 9).

In summary, under comprehensive analysis, the application of a battery is undoubtedly the most suitable solution for using the electrofusion welding machine without a power supply. The TOPSIS evaluation method shows that the use of a Li-ion battery and aluminum alloy material without special constraints is close to the positive ideal solution and furthest from the negative ideal solution. The remaining options corresponding to different innovation scenarios are shown in the Table 6 below. A technology roadmap has been drawn up as shown below (see Fig. 11). The vertical axis represents the type of technology, while the horizontal axis represents time.

The innovative welding machine using lithium batteries combined with aluminium alloys (see Fig. 12). This design is for a welding machine equipped with a battery, which can be adapted to different scenarios by changing the casing material and the type of battery used.

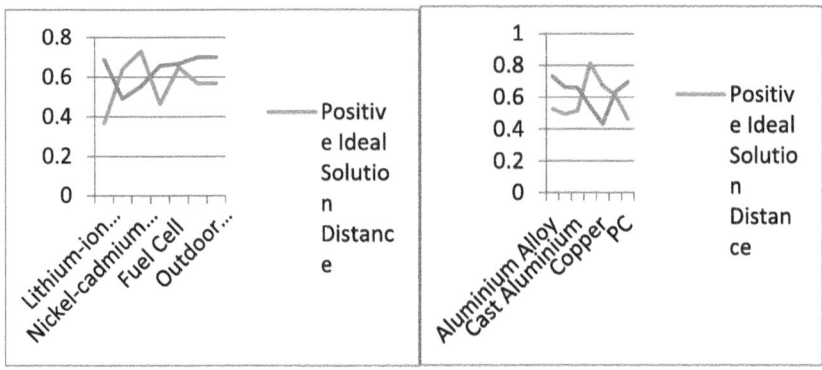

Fig. 10. TOPSIS evaluation results

Table 6. Resources and Counterpart Innovations

Scenarios	Programme
Generic Scenarios	Li-ion battery + aluminium alloy
Low cost factor	Li-ion/NiMH batteries + magnesium alloy
High environmental protection and low pollution	Solar Generator + Aluminium Alloy
Low maintenance cost and high reliability	Lead-acid battery/solar generator + stainless steel
Easy to carry Lightweight	Li-ion Battery + PC/ABS
Low cost	Lead-acid battery/Nickel-cadmium battery + PC/ABS
High power	Outdoor generator + stainless steel

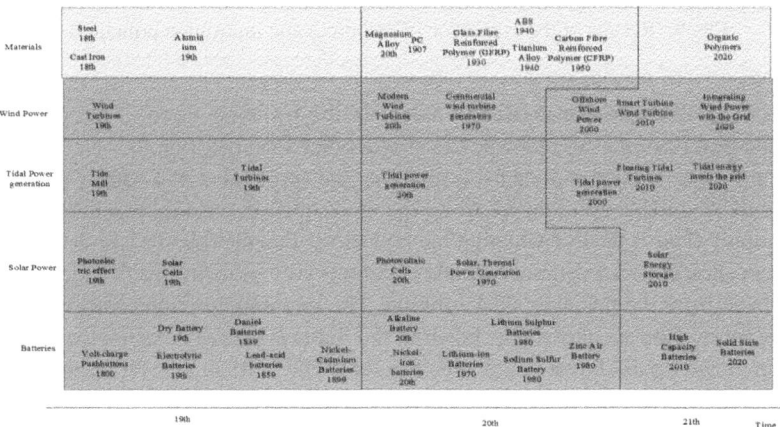

Fig. 11. Technology Roadmap

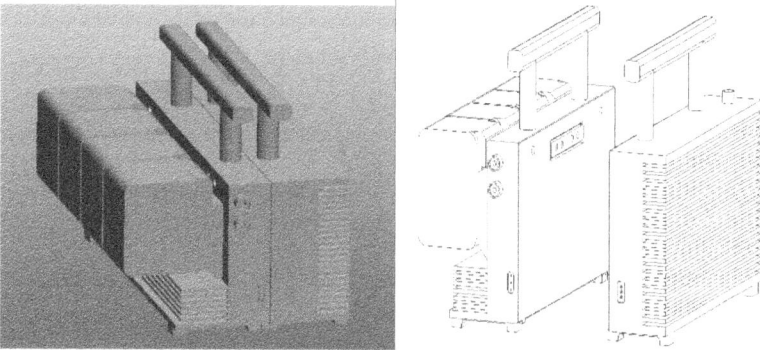

Fig. 12. Electrofusion welding machine

5 Conclusions

This paper takes the product ideal solution as the driving force to excavate the resources needed for product innovation and design, incorporates AI tools to provide assistance in excavating resources, and uses AI to obtain the innovation needs and research hotspots to determine the design direction.

This method has been proven feasible through case studies, combining AI with TRIZ to explore resources and investigating a series of methods for using ChatGPT. The AHP-EWM-TOPSIS evaluation method was constructed to assess the application index of resources in different scenarios. However, this approach still has limitations. For example, the selection of indicator weights relies on expert judgment, and the design of weights varies for different products. This paper only discussed cases based on the use of certain products. Additionally, when applying patent principles from other fields to products with limited patent data, modifications are needed, and they cannot be applied directly. In conclusion, although this method still requires improvement, it provides a foundation and some insights for combining AI and TRIZ in resource exploration.

References

1. Brad, S.: Mapping the evolutionary journey of TRIZ and pioneering its next s-curve in the age of AI-aided invention. In: Cavallucci, D., Livotov, P., Brad, S. (eds.) Towards AI-Aided Invention and Innovation, TFC 2023. IFIP Advances in Information and Communication Technology, vol. 682, pp. 16–36. Springer, Cham (2023). https://doi.org/10.1007/978-3-031-42532-5_1
2. Wessner, J.: Sustainability, TRIZ and packaging. In: Cavallucci, D., Livotov, P., Brad, S. (eds.) Towards AI-Aided Invention and Innovation. TFC 2023. IFIP Advances in Information and Communication Technology, vol. 682, pp. 186–198. Springer, Cham (2023). https://doi.org/10.1007/978-3-031-42532-5_13
3. Li, Z.: Resource Excavation and Application for Technology Innovation. Hebei University of Technology, pp. 1–20 (2017)
4. Yue, Hu.: Research on mining available resources and methods based on TRIZ. Hebei University of Technology, pp. 1–10 (2020)

5. Ma, N.: Research on Multidimensional Resource Extraction and Application Mechanism, Hebei University of Technology, pp. 1–20 (2020)

6. Dong, H.: Reserch on Resource Analysis Method for Conceptual Design of Mechanical Product, pp. 1–15 (2022)

7. Tan, R.: Innovation Design-TRIZ Inventive Problem Solving Theory. Mechanical Engineering Press, Beijing, pp. 33–38, 47–49 (2002)

8. Iouri, B.: Solving problems with Method of the Ideal Result (MIR). In: G Mazur (eds.) Transactions from the 11th Symposium on Quality Function Deployment, pp. 192–203. Novi, Michigan, USA (1999)

9. Liao, Y.: The research of the innovation course and method in product idealization design based on the TRIZ theory. Hunan University, pp. 5–8 (2007)

10. Tom, R., Zwicky, F.: Morphologie and Policy Analysis, pp. 1–4 (1998)

11. Feng, L., Niu, Y., Wang, J.: Development of Morphology Analysis-Based Technology Roadmap Considering Layer Expansion Paths: Application of TRIZ and Text Mining (10), 1–10 (2020)

12. Song, B., Fan, X.: Improved algorithm based on web page features microcomputer applications **23**(1), 18–20 (2002)

13. Brad, S.: Mapping the evolutionary journey of TRIZ and pioneering its next s-curve in the age of AI-aided invention. In: Cavallucci, D., Livotov, P., Brad, S. (eds.) Towards AI-Aided Invention and Innovation. TFC2023. IFIP Advances in Information and Communication Technology, vol. 682, pp. 16–36. Springer, Cham (2023). https://doi.org/10.1007/978-3-031-425 32-5_1

14. Douard, N., Samet, A., Giakos, G., Cavallucci, D.: Navigating the network: how inter-domain information pairing and generative AI can enable rapid problem-solving. In: Cavallucci, D., Livotov, P., Brad, S. (eds.) Towards AI-Aided Invention and Innovation. TFC2023. IFIP Advances in Information and Communication Technology, vol. 682. pp. 152–160. Springer, Cham (2023). https://doi.org/10.1007/978-3-031-42532-5_11

15. Simon, D., Peter, C.: Innovation logic: benefits of a TRIZ-like mind in AI using text analysis of patent literature. In: Cavallucci, D., Livotov, P., Brad, S. (eds.) Towards AI-Aided Invention and Innovation. TFC2023. IFIP Advances in Information and Communication Technology, vol. 682, pp. 108–116. Springer, Cham (2023). https://doi.org/10.1007/978-3-031-42532-5_7

16. Iliass, A., Denis, C.: Multi-domain and heterogeneous data driven innovation problem solving: towards a unified representation framework. In: Cavallucci, D., Livotov, P., Brad, S. (eds.) Towards AI-Aided Invention and Innovation. TFC2023. IFIP Advances in Information and Communication Technology, vol. 682, pp. 140–152. Springer, Cham (2023). https://doi.org/10.1007/978-3-031-42532-5_10

17. Jin, G., Jeong, Y., Yoon, B.: Technology-driven roadmaps for identifying new product/market opportunities: use of text mining and quality function deployment. Adv. Eng. Inform. **29**(1), 126–138 (2015)

18. Liang, Y.: Research on Key Technologies for Innovative Design Based on Patent Mining. Hebei University of Technology, pp. 10–11 (2010)

19. Amir, K.M., Houssin, R., Zouari, A., Dhouib, D., Renaud, J.: Integration TRIZ and MCDM for innovative and sustainable decision-making: a case study in the life cycle of olive oil. In: Cavallucci, D., Livotov, P., Brad, S. (eds.) Towards AI-Aided Invention and Innovation. TFC2023. IFIP Advances in Information and Communication Technology, vol. 682. pp. 237–248. Springer, Cham (2023). https://doi.org/10.1007/978-3-031-42532-5_17

Use of AI in the TRIZ Innovation Process: A TESE-Based Forecast

Oleg Abramov$^{(\boxtimes)}$ (iD)

GEN TRIZ LLC, St. Petersburg, Russia
`oabramov@gen-triz.com`

Abstract. Currently, AI is mainly used in the TRIZ innovation process to find solutions to well-defined technical problems using TRIZ tools such as Inventive Principles, Standards and, occasionally, Functional Oriented Search (FOS). In practice, however, problem solving is usually the least time-consuming part of the innovation process, with most effort normally spent defining the overall goal of the innovation, identifying and analyzing the initial problem, selecting the best solution from the set of solutions found, and justifying its feasibility. Therefore, AI would be much more useful if it were introduced into these labor-intensive parts of the TRIZ innovation process as well, and undoubtedly AI developers will eventually try to automate the entire innovation process. The objectives of this paper are (1) to assess the current effectiveness of using AI in real TRIZ projects, (2) to predict the most likely sequence of future AI implementation in different parts of the TRIZ innovation process, and (3) to identify related challenges. The objectives are achieved by analyzing the composition and timing of various activities in a typical TRIZ project and applying the Trend of Decreasing Human Involvement to these activities, where a TRIZ project is considered a technological process that transforms an initial, poorly formulated problem into a viable solution/product. These results can be used by AI and TRIZ specialists to create a roadmap for integrating AI and TRIZ to produce a fully automated innovation process that is applicable to technical systems and, potentially, to business systems.

Keywords: AI · innovation · TRIZ · TESE · Trends of Engineering Systems Evolution · new product development · NPD

1 Introduction: Objectives of This Research

Artificial Intelligence (AI) today is being actively developed and used in almost all spheres of our life, including innovation, and TRIZ is no exception.

TRIZ has already been used to advance various aspects of AI, e.g., to improve the navigation of mobile autonomous robots using AI [1] or to translate AI-related algorithms from classical to quantum computers [2].

TRIZ developers also actively involve AI in the TRIZ process itself, mainly to facilitate idea generation during the problem-solving phase of the process, when the key problem to solve is clearly defined [3–5]. More details can be found in a recent systematic

© IFIP International Federation for Information Processing 2025
Published by Springer Nature Switzerland AG 2025
D. Cavallucci et al. (Eds.): TFC 2024, IFIP AICT 735, pp. 165–174, 2025.
https://doi.org/10.1007/978-3-031-75919-2_10

literature review by Ghane, Ang, Cavallucci et al. [6], which provides a comprehensive review of advances in integrating AI with TRIZ.

Based on the success of integrating AI with TRIZ, many TRIZ enthusiasts think that the TRIZ process, which usually requires a significant amount of time to learn and use, is now almost fully automated and anyone can instantly find a solution for any problem using some generative AI, such as 'AI TRIZ Master: Solver' [7] or similar.

Although employing AI together with TRIZ to facilitate innovation is undoubtedly a promising trend, the current actual efficacy of using AI in real TRIZ projects seems to be overestimated, as only the least labor-intensive parts of a TRIZ project are currently automated.

Therefore, more effort is needed to integrate AI with various TRIZ tools at all stages of the project.

Thus, the objectives of this research are:

1. To estimate the current practical efficacy of AI use in TRIZ projects,
2. To forecast the future involvement of AI in the TRIZ process, and
3. To identify the main challenges facing the future integration of AI with TRIZ.

The research is limited to the most common types of TRIZ projects: new product/technology development (NPD) and product/technology improvement. Other TRIZ projects, such as those aimed at identifying adjacent markets [8] and technology scouting projects [9], are not considered.

Also, this paper considers only full-fledged TRIZ projects [10] that include various TRIZ analyses of the initial problem and development of fully substantiated solutions. Such projects are routinely performed, for example, by professional TRIZ consultants.

2 Method and Data Used in This Research

The practical effectiveness of using AI in TRIZ projects is assessed based on the potential reduction in resources, i.e., man-weeks, required to perform a project if AI were to be used compared to a non-automated project.

For this purpose, all activities in a TRIZ project and related resources required to perform these activities were considered.

The author used as a baseline a typical TRIZ consulting project, whose structure is similar to that of past projects conducted by him and his colleagues and considered in previous research [11, 12].

Data on the average resources required for various project activities were also taken from these real-life projects. To forecast the sequence of future AI applications in TRIZ projects, the Trend of Decreasing Human Involvement from the Trends of Engineering Systems Evolution (TESE) [13] was used.

Although TESE were first applied to hardware technical systems, the author assumes that this trend can be applied to TRIZ as a data processing system since, as mentioned by Gane et al. [14], they are very broad and versatile and can be employed in a variety of domains.

The main challenges for the future integration of AI with TRIZ were identified based on the literature study and on the author's experience in executing practical TRIZ projects.

3   Results and Discussion

3.1   Current Efficacy of AI Utilization in TRIZ Projects

A typical TRIZ project relating to NPD or product/technology improvement contains 4 stages, each with several activities, as shown in Fig. 1.

As seen in Fig. 1, a professional TRIZ project covers the entire innovation process, beginning with the initial problem (or target problem), which normally is not clearly formulated and often represents more of a business than a technical problem, and ending with the implementation of a specific technical solution that eliminates the target problem.

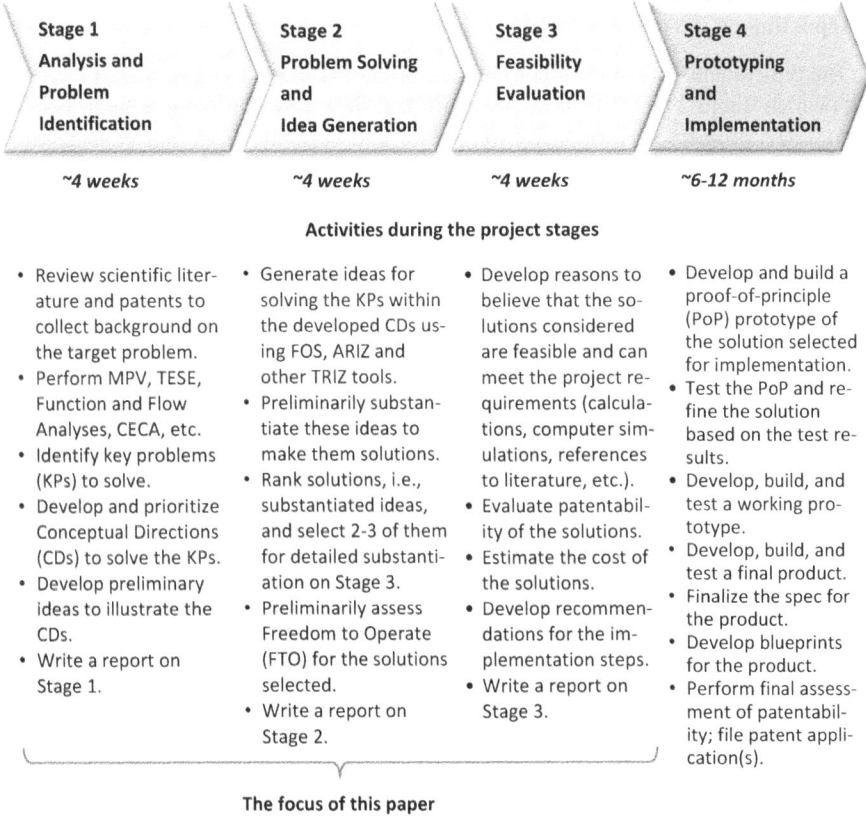

Fig. 1. Example of a typical TRIZ project structure and related activities.

TRIZ is actively used during the first three stages of the project to analyze the target problem, identify underlying key technical problems that must be solved to eliminate the target problem and generate solutions for these key problems. During Stage 4, however, TRIZ is not often employed. Instead, developers usually rely on regular engineering tools such as various CADs to develop, build and test all necessary prototypes and, eventually, produce the final product or technology.

This paper is, therefore, focused on the first three stages of a TRIZ project, where TRIZ is most needed.

All activities in Stages 1–3 shown in Fig. 1 can be attributed to one of seven types:

1. **Literature review** of scientific papers and patents
2. **TRIZ analysis:** Main Parameters of Value (MPV), TESE, Function and Flow analyses, Cause and Effect Chains Analysis (CECA), etc.
3. **Idea generation** using TRIZ tools such as Function-Oriented Search (FOS) and ARIZ
4. **Idea substantiation** to prove technical and economic viability of ideas
5. **Benchmarking** (ranking) of ideas and solutions (substantiated ideas)
6. **Patent analysis** to evaluate the freedom to operate and novelty of solutions
7. **Reporting** on the findings of each stage

The typical amount of resources used for each of the above-listed project activities is shown in Table 1 below. These are empirical data obtained from a large pool of actual TRIZ projects performed by the author and his colleagues and considered in their previous research [11, 12]. Here these data are used as a baseline for assessing the potential effectiveness of AI in TRIZ projects.

Table 1. Resources used for different activities in non-automated TRIZ projects.

Project activities	Resources used	
	man-weeks	%
Literature review	5	14
TRIZ analysis	6	17
Idea generation	4	11
Idea substantiation	8	22
Benchmarking	2	6
Patent analysis	3	8
Reporting	8	22
Total:	36 man-weeks	100%

Many existing AI tools already make it possible to automate some of the TRIZ project activities. For example:

- TRIZ and Generative AI-V3.0 [4], AutoTRIZ [5] and AI TRIZ Master: Solver [7] are claimed to generate ideas if the user provides a prompt with a clearly stated problem.
- Semantic TRIZ (S-TRIZ) [6] and PatentInspiration tools [15, 16] can be used to facilitate patent analysis.
- SciSpace [17] can, to some extent, help to automate a literature review.

Although some people may think these AI tools fully eliminate the human labor required to perform these activities, all these tools require constant interactive communication with the user, who develops and adjusts the prompts and evaluates the AI-generated outputs. This means that AI tools do consume significant resources.

Unfortunately, the efficacy of using AI tools in TRIZ projects to reduce the human labor used in a project is not indicated in any of the papers found by the author.

Therefore, in this study we used two values for assessing the potential efficacy of existing AI-powered TRIZ tools:

1. The theoretical maximum, calculated as if the tool fully automated the related project activity and no human labor is needed at all
2. The anticipated maximum practical efficacy, calculated as half of the theoretical maximum – to consider the necessity of some human labor for creating prompts, etc.

The results of this assessment are given in Table 2. As seen from this table, the tools, even if used together, could currently reduce the total amount of resources needed for a TRIZ project or accelerate the project only by about 16%, which is a good, but not very impressive number.

Unfortunately, the activities that require the most time and resources - TRIZ analysis, idea substantiation and reporting - have not in fact been automated yet. To accelerate TRIZ-based innovations significantly, these activities should be automated.

Table 2. Potential reduction of resources used for different activities in an automated TRIZ project compared to a non-automated project.

Project activities	Reference to AI tools that can be used (examples)	Potential reduction of resources used, %	
		Theoretical Max	Practical Max
Literature review	SciSpace [17]	14	7
TRIZ analysis	Not found	-	-
Idea generation	AutoTRIZ [5]	11	5.5
Idea substantiation	Not found	-	-
Benchmarking	Not found	-	-
Patent analysis	PatentInspiration [15, 16]	8	4
Reporting	Not found	-	-
Total:		33%	16.5%

It must be noted that a literature review [6] indicates that a significant number of papers devoted to AI in TRIZ relate to analytical TRIZ tools. However, these tools are limited to Component and Function analyses (26% of papers) and TESE (13%) while the integration of AI with the most time-consuming analytical tools such as MPV analysis [18] and CECA [19] is not considered in the papers.

In addition, some works on the use of AI in TRIZ analytical tools simply explore the benefits and challenges of integrating these tools with AI. For example, Ghane et al. [14] explore the potential benefits and challenges of using AI for TESE analysis.

Therefore, AI cannot at present significantly accelerate TRIZ analysis.

3.2   Future Involvement of AI in the TRIZ Process

The integration of AI with TRIZ will undoubtedly continue and is expected to start a new S-curve for TRIZ [20]. This process corresponds to the Trend of Decreasing Human Involvement, according to which the functions performed by humans in a technical system or in a technological process are gradually automated in the following order [13]:

1. Main function
2. Transmission functions
3. Control functions
4. Decision making

A TRIZ project (Fig. 1) can be considered as a technological process with the target problem being input data and the new or improved old product being the output.

Correspondingly, Stages 1–3 of the project represent a technological process that yields the best technical solutions to be further prototyped and implemented in a product, as shown in Fig. 2.

In the process shown in Fig. 2, the main functions that, according to the Trend of Decreasing Human Involvement, should be automated first, refer to the core functionality of individual TRIZ tools. As shown in the previous section, some of the problem-solving tools are already being automated by AI while TRIZ analytical tools and substantiation activities remain unautomated.

The transmission functions in this process that should be automated next include the transfer of information from one project activity to another, i.e., from one TRIZ tool to another. To date, AI does not seem to be used to automate these functions.

Control functions, which are related to building an optimal project roadmap that is created using an optimal set of TRIZ tools for a specific project, should be automated after transmission functions. This part of the TRIZ innovation process is also not yet automated.

Decision making in the process shown in Fig. 2, which should be automated last, refers to selecting the target problem to be addressed in the project, selecting the right MPV to implement in a new product (this is part of MPV analysis) and selecting the best solution to develop a new product (part of benchmarking). All these activities are currently performed by humans.

3.3   Challenges to the Future Integration of AI with TRIZ

Although the integration of AI with TRIZ is inevitable, it will not be a simple process.

For example, the following barriers for using AI in product development are mentioned by Müller, Roth and Kreimeyer [21]:

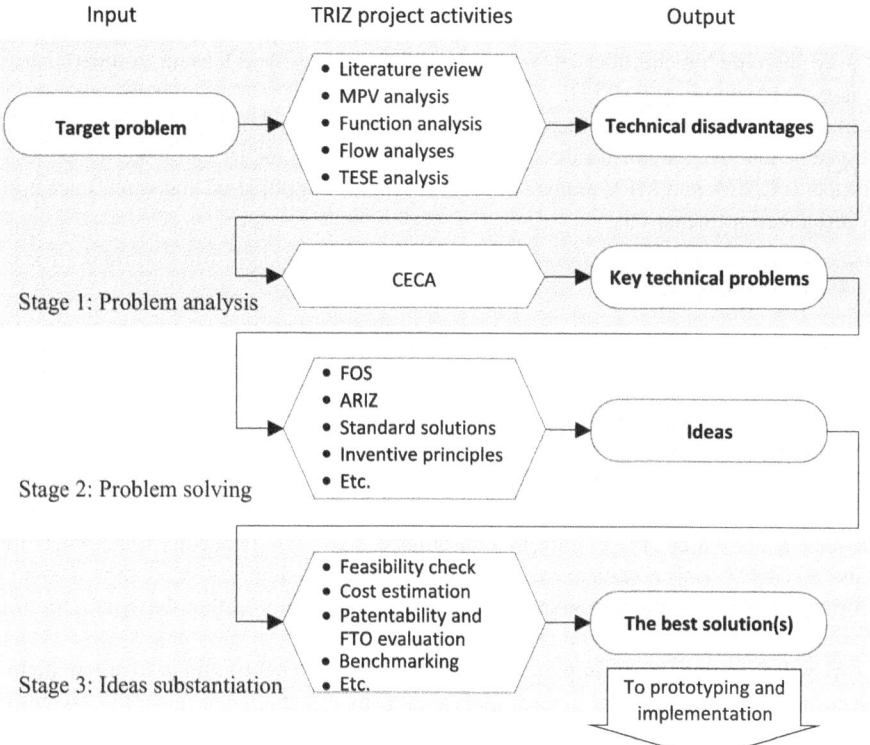

Fig. 2. TRIZ project (Stages 1–3) as a technological process: simplified illustration.

- The scientific meta models do not provide enough information to develop AI applications due to their low level of detail.
- There is no standardized method enabling companies to collect and describe the available data in the context of their development activities.
- There is a lack of standardized process models for AI implementation.

In Fig. 1, these barriers mostly relate to Stage 4 of the project, but they also, to some degree, affect Stage 3.

Further barriers to using AI in the first three stages of the project can be added to the list, for instance:

- The results produced by AI are highly dependent on the prompts used, suggesting iterative communication between humans and the AI to adjust the prompt. This makes it difficult to fully eliminate human involvement in the process.
- Lack of integration of existing TRIZ AI with CAD and other engineering tools, as well as the overly general solutions suggested by the AI, makes it difficult to automate solution substantiation at Stage 3 of the project.
- Sensitive information such as prompts used to solve a problem as well as solutions found by the AI could potentially be exposed to third parties [22], which makes companies reluctant to use AI to solve their problems.

- AI may generate false results, which is unacceptable in a TRIZ project as it may increase the number of wrong decisions made by AI users compared to a non-automated process [23].
- In the author's experience, there is a certain amount of tacit knowledge in any technical system that AI cannot access. This can make it difficult to automate TRIZ tools such as CECA and MPV analysis, which often use unpublished and sometimes non-verbalized information.

4   Conclusions

Despite some success in integrating AI with TRIZ tools and a great enthusiasm for this among many TRIZ professionals, the effectiveness of using AI is not currently very high, and the potential practical reduction of a TRIZ project's duration by using AI does not exceed 16%.

To dramatically increase the efficacy of using AI in TRIZ, it is necessary to automate the most time-consuming activities in a TRIZ project such as TRIZ analysis and Idea Substantiation, which are not currently automated. First, it is necessary to automate the individual TRIZ tools associated with these activities, and then, automate the transfer of information between the various tools within these activities so that these activities do not require human participation at all.

AI will then need to be implemented to develop a project roadmap to automate the selection of TRIZ tools used in each individual project, which can make the execution of the entire project almost completely automated.

Finally, the AI should be able to make decisions such as: identify the target problem that the project should solve, select which MPVs the new product should address, and what best solution should be implemented in the new product.

Examples of key challenges that need to be addressed for widespread use of AI-based professional TRIZ tools include integrating these tools with CAD, improving the reliability of AI results to eliminate the possibility of false results, providing sufficient security for sensitive information used in a project, and developing ways for AI to access the tacit knowledge needed to perform exhaustive CECA and MPV analysis.

References

1. Florian, A.V., Brad, S.: TRIZ driven identification of AI application to improve navigation of mobile autonomous robots. In: Cavallucci, D., Brad, S., Livotov, P. (eds.) Systematic Complex Problem Solving in the Age of Digitalization and Open Innovation. TFC 2020. IFIP Advances in Information and Communication Technology, vol. 597, pp. 15–29. Springer, Cham (2020). https://doi.org/10.1007/978-3-030-61295-5_2
2. Guehika, A.: TRIZ, a systematic approach to create quantum activation function for deep learning's hidden layers, in order to make AI explainable with quantum computer. In: Benmoussa, R., De Guio, R., Dubois, S., Koziołek, S. (eds.) New Opportunities for Innovation Breakthroughs for Developing Countries and Emerging Economies. TFC 2019. IFIP Advances in Information and Communication Technology, vol. 572, pp. 371–387. Springer, Cham (2019). https://doi.org/10.1007/978-3-030-32497-1_30

3. Brad, S., Ștetco, E.: An interactive artificial intelligence system for inventive problem-solving. In: Nowak, R., Chrząszcz, J., Brad, S. (eds.) Systematic Innovation Partnerships with Artificial Intelligence and Information Technology. TFC 2022. IFIP Advances in Information and Communication Technology, vol. 655, pp. 165–177. Springer, Cham (2022). https://doi.org/10.1007/978-3-031-17288-5_15

4. Pheunghua, T., Adunka, R.: TRIZ and Generative AI-V3.0. ResearchGate (2024). https://www.researchgate.net/publication/380036640_TRIZ_and_Generative_AI-V30. Accessed 24 May 2024

5. Jiang, S., Luo, J.: AutoTRIZ: Artificial Ideation with TRIZ and Large Language Models. arXiv preprint, arXiv:2403.13002v2 [cs.HC] (2024)

6. Ghane, M., Ang, M.C., Cavallucci, D., Kadir, R.A., Ng, K.W., Sorooshian, S.: Semantic TRIZ feasibility in technology development, innovation, and production: a systematic review. Heliyon 10(1), e23775 (2024)

7. QM&E Innovation. AI TRIZ Master: Solver. [AI-based software tool]. https://gptstore.ai/gpts/oIOFM9fS5o-ai-triz-master-solver. Accessed 04 June 2024

8. Abramov, O., Medvedev, A., Tomashevskaya, N.: TRIZ roadmap for identifying adjacent markets. In: Proceedings of the 15th MATRIZ TRIZfest-2019 International Conference, pp. 382–390. Heilbronn, Germany (2019)

9. Fimin, P., Smirnova, E., Abramov, O.: Technology Scouting for Solid Waste Treatment: TRIZ Approach. Acta Technica Napocensis, Series: Applied Mathematics, Mechanics, and Engineering, vol. 63, Special Issue (2020)

10. Abramov, O.: TRIZ-assisted stage-gate process for developing new products. J. Financ. Econ. 2(5), 178–184 (2014)

11. Abramov, O., Medvedev, A., Rychagov, V.: Evaluation of the effectiveness of modern TRIZ based on practical results in new product development. In: Benmoussa, R., De Guio, R., Dubois, S., Koziołek, S. (eds.) New Opportunities for Innovation Breakthroughs for Developing Countries and Emerging Economies. TFC 2019. IFIP Advances in Information and Communication Technology, vol. 572, pp. 36–44. Springer, Cham (2019). https://doi.org/10.1007/978-3-030-32497-1_4

12. Abramov, O., Tomashevskaya, N., Medvedev, A., Rumiantsev, K.: Who needs TRIZ consultants: a case study. In: Proceedings of the 1st International TRIZ Conference 2023 (ITC-2023), Graz, Austria, pp. 125–133 (2023)

13. Lyubomirskiy, A., Litvin, S., Ikovenko, S., Thurnes, C.M., Adunka, R.: Trends of Engineering System Evolution (TESE): TRIZ Paths to Innovation. TRIZ Consulting Group GmbH (2018)

14. Ghane, M., Ang, M.C., Cavallucci, D., Kadir, R.A., Ng, K.W., Sorooshian, S.: TRIZ trend of engineering system evolution: a review on applications, benefits, challenges and enhancement with computer-aided aspects. Comput. Ind. Eng. 174, 108833 (2022)

15. Dewulf, S., Childs, P.R.N.: Innovation logic: benefits of a TRIZ-like mind in AI using text analysis of patent literature. In: Cavallucci, D., Livotov, P., Brad, S. (eds.) Towards AI-Aided Invention and Innovation. TFC 2023. IFIP Advances in Information and Communication Technology, vol. 682, pp. 95–102. Springer, Cham (2023). https://doi.org/10.1007/978-3-031-42532-5_7

16. AULIVE. PatentInspiration. [AI-based software tool]. https://www.patentinspiration.com. Accessed 05 June 2024

17. Typeset. SciSpace. [AI-based software tool]. https://typeset.io. Accessed 04 June 2024

18. Abramov, O.: Product-oriented MPV analysis to identify voice of the product. In: Proceedings of the 12th MATRIZ TRIZFest-2016 International Conference, Beijing, People's Republic of China, pp. 231–243 (2016)

19. Abramov, O.: TRIZ-based cause and effect chains analysis vs root cause analysis. In: Proceedings of the 11th MATRIZ TRIZFest-2015 International Conference, Seoul, South Korea, pp. 283–291 (2015)

20. Brad, S.: Mapping the evolutionary journey of TRIZ and pioneering its next S-Curve in the age of AI-aided invention. In: Cavallucci, D., Livotov, P., Brad, S. (eds.) Towards AI-Aided Invention and Innovation, TFC 2023. IFIP Advances in Information and Communication Technology, vol. 682, pp. 3–22. Springer, Cham (2023). https://doi.org/10.1007/978-3-031-42532-5_1

21. Müller, B., Roth, D., Kreimeyer, M.: Barriers to the use of artificial intelligence in the product development – a survey of dimensions involved. In: Proceedings of the International Conference on Engineering Design (ICED 2023), Bordeaux, France, pp. 757–766 (2023)

22. Daniels, J.: How Generative AI Can Affect Your Business' Data Privacy. Forbes. https://www.forbes.com/sites/forbesbusinesscouncil/2023/05/01/how-generative-ai-can-affect-your-business-data-privacy/. Accessed 11 June 2024

23. Agudo, U., Liberal, K.G., Arrese, M. Matute, H.: The impact of AI errors in a human-in-the-loop process. Cogn. Res. Princ. Implic. **9**(1), (2024)

Evaluating the Effectiveness of Generative AI in TRIZ: A Comparative Case Study

Nikhil Phadnis$^{(\boxtimes)}$ ⓘ and Marko Torkkeli

Lappeenranta-Lahti University of Technology, 53850 Lappeenranta, Finland
{nikhil.phadnis,Marko.torkkeli}@lut.fi

Abstract. The rapid advances in generative AI technologies have sparked a debate among researchers on their role in the innovation process, particularly regarding their problem-solving and idea-generation capabilities. While researchers theorise the potential of generative AI in conjunction with TRIZ (Theory of Inventive Problem Solving), evaluating its current state and understanding its practicality is equally critical. Hence, this paper provides evidence of generative AI's ability to offer solutions in real innovation projects. Our exploratory study compares the results of an actual innovation project in a professional consulting-like setting using traditionally applied modern TRIZ tools against generative AI-assisted results for the same customer-defined problem. The comparison focuses on the solutions' degree of similarity, depth, and breadth. Additionally, our research identifies the advantages, disadvantages, and feasibility of using generative AI in problem-solving and innovation projects. Our findings indicate that combining generative AI and TRIZ produces feasible, cross-domain preliminary conceptual directions with satisfactory scientific substantiation. Lastly, we recommend suitable use cases for innovation managers and TRIZ practitioners, highlighting how the TRIZ-GPT combination can save considerable time exploring preliminary concepts and idea generation during problem-solving.

Keywords: Generative AI · TRIZ · Case study · GPT

1 Introduction

Past research on the theory of inventive problem-solving (TRIZ) has had an interesting evolutionary methodical transformation from a prominent problem-solving methodology to a globally recognised innovation-supporting toolkit and finally to an integrated part of broader innovation practices across several domains [1]. Fuller et al. state that artificial intelligence (AI) will play an essential role in the innovation process, from exploring problems to selecting solutions [2]. The role of AI is expected to augment human creativity and automate specific parts of the problem-solving process. Other researchers also forecast that AI will become the key driver of innovation management, leading to more open and collaborative approaches and the emergence of new roles in innovation teams [3]. A recent evolutionary study by Brad, 2023 on the TRIZ methodology predicts that computer science, AI and its integration with TRIZ will be the next

© IFIP International Federation for Information Processing 2025
Published by Springer Nature Switzerland AG 2025
D. Cavallucci et al. (Eds.): TFC 2024, IFIP AICT 735, pp. 175–192, 2025.
https://doi.org/10.1007/978-3-031-75919-2_11

breakthrough in the problem-solving method. Moreover, the research also suggests that TRIZ will supposedly undergo seamless integration and rapid adoption due to the fast pace of artificial intelligence development [4].

While there has been an increase in academic publications on AI and TRIZ and thematic conferences inviting scientific journals, there remains a notable gap in research on the generative pre-trained transformer (GPT) AI models and TRIZ. The most recent study on TRIZ and GPT by Pheunguha and Adunka, 2024 focuses on prompt engineering for GPT technologies, applied in conjunction with TRIZ tools for better results. They attempt to address the issue of subjective prompts that lead to the high variability of results. Their work provides a systematic handbook that uses different structures of prompts engineered for specific purposes to enhance the problem-solving process [5]. However, they do not provide evidence of case studies showcasing their effectiveness in real projects. In this research, we aim to address this gap by providing evidence of the combination of TRIZ and generative AI and evaluating its practical applicability.

Therefore, in our exploratory study, we compare the results of generative AI chatbots with TRIZ against manually applied GENTRIZ methodology to solve the same customer-defined problem wherein the results were delivered to a client in a professional consulting-like environment [6]. We aim to determine the degree of alignment between the solutions delivered to the client and those generated by AI, demonstrating its feasibility, accuracy and robustness in real-world applications. This raises the following research question: 'Can the TRIZ-ChatGPT combination generate practical and usable solutions in real projects, potentially replacing modern TRIZ methods?'.

2 Methodology

The research method used to select the case study is purposive sampling. The researchers selected this method because of its convenience and ease of data availability [7]. Additionally, the nature of the problem involves a limited number of system and supersystem components that make it easy for a broad audience to understand. The methodology is divided into three sections: the case background, an overview of the GENTRIZ methodology, and the AI methodology used in the research.

2.1 Case Study Background

In this research, we selected a project in a professional consulting service-like environment completed in 2020 by a qualified TRIZ practitioner and his team. We made sure that the same researcher also performed the AI-assisted TRIZ later to standardise human bias performed in both cases. The case involves a healthcare and life sciences company that produced pipettes and pipette racks for laboratory applications. The project aimed to "increase the sustainability of their pipette racks". Pipettes are small laboratory instruments typically used in chemistry or biology to move and measure liquids. Often, they have a removable attachment known as a pipette tip. The plastic box that holds the pipette tips is known as a pipette rack. Figure 1 below shows the pipette tips and the rack we refer to. As a part of the sustainability transition, the company switched its plastic material from polypropylene (PP) to fully recyclable thermoplastic polyethene terephthalate (PET), which resulted in less plastic and weight of the rack. The case, as described by the customer, is as follows.

"As a standard operating procedure, researchers and lab technicians are instructed to clean and sterilise the pipette racks in an autoclave machine at recurring intervals. An autoclave uses steam under pressure to kill harmful bacteria, viruses, and spores to sterilise the pipette rack at 120 °C for approximately 30 min. The original material, PP, was thermally resistant to the autoclave machine; however, the problem occurred when the new recyclable material, PET, could not withstand the thermal load in the autoclave and crumpled due to heat. PET had inferior thermal resistance and functional and structural properties that did not meet the requirements. Therefore, PET racks became consumable and unsuitable for repeated use since they couldn't be sterilised. Thereby reducing its usable life significantly.

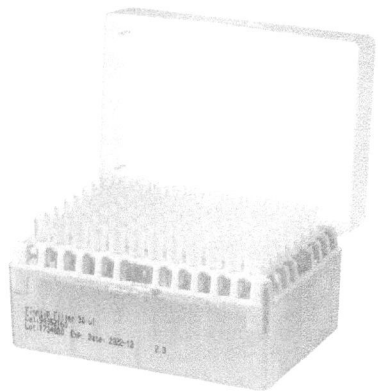

Fig. 1. Representative figure of Pipette racks with Pipette tips

Project Goal: Find alternative materials for pipette tip racks within the constraints of the following requirements.

1. Can be produced in the same or similar form factor to the current racks
2. Low-cost
3. Robust- capable of support tips during loading and withstand shipping and handling
4. Clean and sterilisable (no contamination on the tips)- steam autoclave built for 20 min at 120 °C for ten cycles and gamma radiation resistant
5. Chemical compatibility similar to PP."

The company's original problem statement suggests finding alternative materials as the direction of the solution.

2.2   Overview of the Application of Modern TRIZ on the Case.

This section summarises the manually applied GENTRIZ approach, project logic, TRIZ tools applied, key problems, and conceptual directions for the case employed by the TRIZ team. The section also briefly overviews the key problems and their emerging conceptual directions. Due to confidentiality reasons, the researchers decided not to reveal any detailed solutions in this study. Several iterative discussions with the company led to the scope refinement and allowed the TRIZ practitioners to transform the initial problem into key problems by applying TRIZ tools. The project objective was refined to "identify cost-effective technical solutions that allow existing recyclable racks to be autoclavable."

The TRIZ team benchmarked existing systems and performed preliminary patent landscaping, function analysis, and cause-and-effect chain analysis for the problem identification stage, resulting in key disadvantages and key problems in line with the GENTRIZ methods [6]. This is followed by the problem-solving stage, which includes generating conceptual directions and concept substantiation. Figure 2 illustrates the project logic followed by the TRIZ team using the GENTRIZ approach.

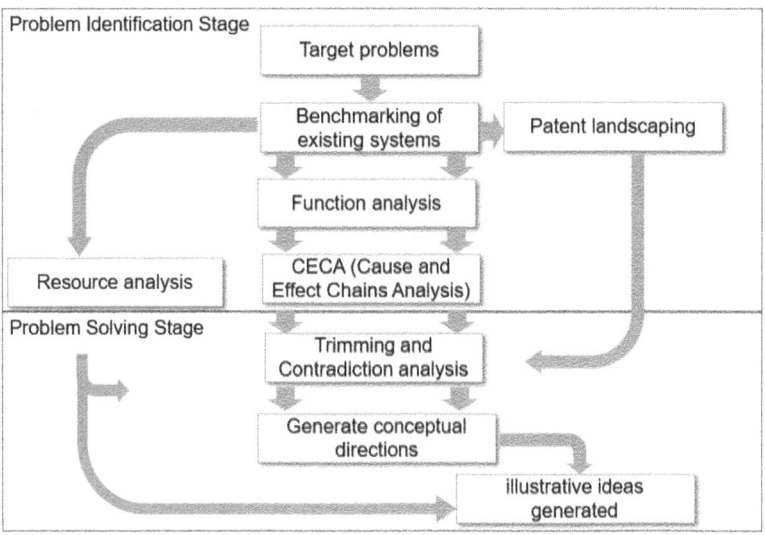

Fig. 2. Project logic with GENTRIZ approach [6]

In this case, trimming was not performed since the scope to trim the component was limited. Figure 3 illustrates the key problems and their respective general conceptual directions in which specific solutions were developed. However, the researchers have omitted specific solutions in the conceptual directions for reasons of confidentiality.

Five key problems were revealed during the problem identification stage. Conceptual direction one (CD1) was out of scope since changing sterilisation machines is a super-system element that the company did not control and was not pursued in detail. The analysis found three key problems from cause-and-effect chain analysis and one key problem from contradiction analysis. Post the problem identification stage, the researchers formulated several contradictions for each conceptual direction and developed solutions that allowed the company to meet its requirements. Some solutions recommended alternative materials as substitutes whilst meeting all the customer requirements. Several solutions were directed towards better functional strength and dynamisation of the pipette racks,

#	Key Problem	Conceptual direction
1	How to eliminate Key Disadvantage "Excessive Temperature of steam"?	CD1: Alternate methods for Sterilization or changing autoclave mechanisms
2	How to eliminate Key Disadvantage "Insufficient Thickness of the material"? How to increase "Glass transition temperature" of plastic?	CD2: Creating Stronger Materials/structures and alternate materials
3	"Rack Should be thin In order to have low weight BUT Rack Should be thick In order to resist heat"?	CD3: Dynamization of the rack
4	How to eliminate Key Disadvantage "Material Holds heat"?	CD4: Movement of heat
5	How to eliminate Key Disadvantage "steam contacts rack directly"?	

Out of scope directions	From Cause-and-Effect Chain Analysis	From Contradiction Analysis

Fig. 3. Key problems and conceptual directions formulated by the TRIZ team

resolving the contradiction in key problem four. Lastly, a group of solutions was directed towards moving heat away from the plastic in the operation zone during the operation time, which required minimal changes to the existing racks. In total, the TRIZ team delivered 20 + promising solutions offered to the client. We will use these delivered solutions to compare them with the AI-generated solutions.

2.3 AI Methodology and Assumptions

The section elaborates on the artificial intelligence-related methodology followed during the research. Pheunguha and Adunka highlight the importance of prompt engineering in their study and summarise six different prompt design techniques that users can employ for better GPT results [5]. Similarly, in our case, the input prompts are incredibly critical for productive output. However, given the human subjectivity, the input prompts provided to the GPT will be highly subjective and variable. Therefore, to resolve this problem, we used a commercially available Advanced Innovation Design Approach (AIDA) automatic idea and IP generator version 2.09 software to create standardised and consistent prompts [8, 9]. The AIDA software is commercially available and developed by Tris Europe Innovation Academy, founded by Professor Pavel Livotov and his team. This Microsoft Excel-based software provides a feature for generating a prompt structure that can be copied and pasted into any generative AI-powered chatbots without needing any iterative prompt engineering. The software was created primarily for audiences with TRIZ knowledge, which brings us to our first assumption: users should have a basic understanding of TRIZ to use the AIDA software effectively. Figure 4 shows the landing page of the AIDA software and its AI prompt generation feature. In this case, we selected AI prompter 2, specifically utilised for developing product concepts.

The software uses a specific structure to model any problem description into a pre-defined prompt that does not require constant human refinement. We will describe each step briefly and provide justifications for how it was used for this research.

Step 1. The user inputs the context of the problem or the invention task
Our research aims to determine if generative AI can resolve the problem described in the

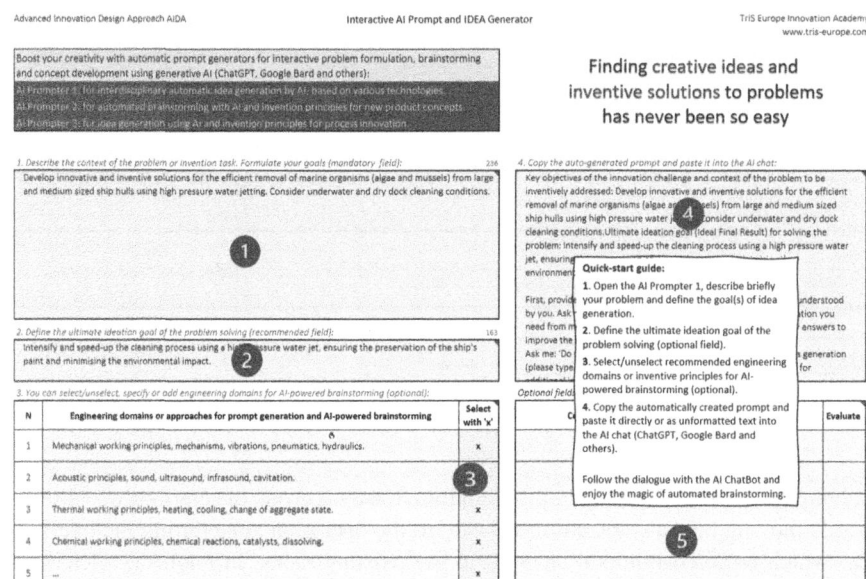

Fig. 4. Screenshot of AIDA software V2.09 interface and steps in the prompt generation (Tris-Europe Innovation Academy)

company's initial problem description. Therefore, we used the company's unchanged description and entered it into the prompt generator. The prompt was the same as described in the case study background section and the formulated goals below.

"As a standard operating procedure, researchers and lab technicians are instructed to clean and sterilise the pipette racks in an autoclave machine at recurring intervals. An autoclave uses steam under pressure to kill harmful bacteria, viruses, and spores to sterilise the pipette rack at 120 °C for 30 min. The original material, PP, was thermally resistant to the autoclave machine; however, the problem occurred when the new recyclable material PET could not withstand the thermal load in the autoclave and crumpled due to heat. PET material also had inferior thermal resistance and functional and structural properties that did not meet the requirements. Therefore, PET racks became consumable and unsuitable for repeated use since they couldn't be sterilised. Thereby reducing its usable life.

Find alternative materials for pipette tip racks within the constraints of the following requirements. 1. Can be produced in the same or similar form factor to the current racks 2. Low-cost 3. Robust- capable of providing support tips during loading and withstanding shipping and handling. 4. Clean and sterilisable (no contamination on the tips)- steam autoclave built for 20 min at 120 °C for 10 cycles and gamma radiation resistant 5. Chemical compatibility similar to PP."

Step 2. The user describes the ultimate ideation goal of problem-solving
This is an optional but recommended field for the user to fill in in the AIDA software and is also known as an ideal final result (IFR), wherein some examples provided by the software follow the structure of "improve the useful action XXX, eliminate the harmful

effect XXX, resolve the contradiction XXX". This step requires the user to demonstrate basic TRIZ knowledge and formulate an IFR. Therefore, we provided the most suitable IFR for the initial problem in the recommended format: Eliminate the Harmful effect of "steam melting the PET pipette rack."

Step 3. The user is asked to select/unselect inventive solution principles for AI-powered brainstorming

This step is optional but important in the solutions generated. The AIDA software consists of several elementary solution principles and applies them to the working tool, target object, workpiece affected by the working tool, to the main function or the undesired function as identified in steps 1 and 2 [10]. These elementary solution principles go well beyond the forty inventive principles. Eighty-eight such elementary solution principles exist; however, we left it in its default condition with ten elementary solution principles selected. Some examples of the solution principles described are "use Work Tool in other aggregate states: change it to solid, liquid, gas, or plasma." Most elementary solution principles are variations and structured specifically for the working tool. Therefore, human oversight over generative AI is critical to validate whether the working tool and the target object were identified appropriately.

Step 4. Copying the auto-generated prompt

Next, we copy the single prompt automatically generated in the Excel sheet and paste it into the Open AI ChatGPT 4o chatbot. We noticed a discrepancy concerning this step that can create a variation in outcomes. We notice that the overall output of the conceptual directions remains similar, with minor but acceptable differences. In some cases, while testing for repeatability, the generative AI fails to understand and follow the instructions, making it inconvenient for the user. Therefore, the user must repeat this step. Lastly, step five is a feature in the AIDA software that allows users to take notes of ideas in the software.

2.4 Selection of the GPT Model and AIDA Prompt Generator

This section justifies the generative AI model selected and the prompt output structure. We selected ChatGPT 4o, OpenAI's latest state-of-the-art flagship model behind the generative AI-powered chatbot capable of image recognition and generation, the latest available model as of July 2024. It provided reliable and more repeatable results than Google Gemini and Anthropic Claude. Gemini and Claude produced outcomes that took an excessively creative stance and were not considered. However, it was noted that GPT4o provides the most comprehensive and consistent response, following the user instructions on repeated trials of the prompt. All chat history has been recorded and saved for research purposes using the link in the appendix section.

Figure 5 illustrates the GPT4o logic diagram and its interaction, highlighting the key stages where user input is required. The logic below directly corresponds to the AIDA software prompt and its working principle. Problem-solving with ChatGPT happens in two distinct stages. Firstly, the user is asked to refine the prompt iteratively and provide an output to confirm and verify the initial problem. Moreover, GPT can ask relevant additional questions to clarify the context and request user input. In this stage, the structure of the AIDA software forces ChatGPT to define the initial problem situation and the working tool. It also outputs the main useful function, the target object and the user's desired IFR. If the user confirms and verifies all the above, the GPT provides a revised, improved output and requests the user to proceed to the idea generation stage.

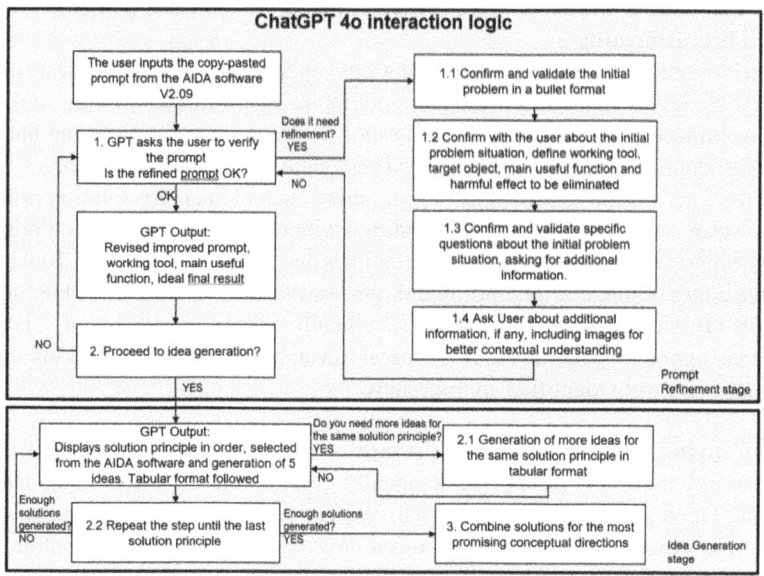

Fig. 5. ChatGPT 4o logic interacting with the user input

In the idea generation stage, user-predefined solution principles from the AIDA prompt are used to generate the top five ideas in a tabular format for each selected solution principle. Note that these are not inventive principles in TRIZ but solution principles. At its core, they are segmented into specific variations and versions of inventive principles that mostly apply to the working tool. Ten solution principles specified in the AIDA software were selected by default; therefore, at least fifty ideas will be generated. However, TRIZ practitioners can choose the most relevant solution principles based on the description provided in the AIDA software, which modifies the initial prompt and, therefore, the output of GPT. During the idea generation process, the GPT provides the idea name, ID description, implementation example (some scientific justification) and the solution principle used in the idea. If the user is satisfied with the number of ideas generated, it moves on to the following principle. Post-idea generation, the AIDA software consists of a prompt that combines several ideas and provides integrated complementary ideas to provide robust and effective solutions. Lastly, ChatGPT generates and displays five complementary integrated conceptual directions in a tabular format following the structure.

3   Results

The results present ChatGPT 4o generated 50 ideas and five integrated conceptual descriptions that we will compare against the solutions to the ones delivered by the TRIZ team. Firstly, Table 1 shows the top five conceptual directions that ChatGPT generated. The table shows the AI-generated concept name, conceptual description, idea ID numbers, and implementation examples.

As shown in Table 1, the AI output typically consists of the conceptual description, which provides the general action principle. For example, CD1 and its AI-generated concept description suggest materialistic changes. This conceptual direction is aligned with material science and polymer chemistry action principles. Next, the AI result provides an implementation example, which dives deeper and provides more specific answers. For instance, high-performance temperature-resistant polymers, such as PEEK, are the specific material for use with some scientific justification. We added the similarity and depth column to compare the AI-generated and the manual results. The similarity column marks a "Yes" if the solutions generated by AI are similar or the same as compared to the GENTRIZ method and a "No" if found otherwise. Next, the depth column marks a "yes" if the solution description and implementation example have sufficient details to develop a feasible solution for the problem solver and a "no" if found otherwise.

Table 1. The AI-generated result in comparison to results from the TRIZ team

CD	AI- Concept Name	AI-Concept Description	AI- Implementation Example	Similarity	Depth
1	Composite Material Approach	Utilise a high-temperature composite material combining PET with other polymers and granules to enhance thermal resistance and structural integrity	Develop a pipette rack using a blend of PET and high-temperature-resistant polymers like PEEK, reinforced with thermally resistant granules such as silica or alumina, to create a composite material that with-stands autoclaving without deformation	Yes	Yes
2	Modular and Dismountable Design	Design the pipette rack with inter-changeable and dismountable modules that can be easily assembled and disassembled for cleaning and autoclaving	Create a pipette rack consisting of modular, snap-together parts that can be assembled into smaller components for autoclaving and reassembling afterwards. If damaged, each part can be individually replaced, reducing waste and cost	No	No

(continued)

Table 1. (*continued*)

CD	AI- Concept Name	AI-Concept Description	AI- Implementation Example	Similarity	Depth
3	Advanced Coating and Surface Treatments	Apply various coatings and surface treatments to enhance PET racks' thermal resistance and durability during autoclaving	Coat the PET racks with nanocomposite materials and thermal-resistant chemical layers, such as silicone or ceramic-based coatings, to protect the underlying PET from heat damage and improve overall thermal stability	No	No
4	Phase Change and Cooling Integration	Incorporate phase change materials and cooling inserts within the rack design to absorb excess heat during autoclaving and prevent deformation	Integrate phase change materials that absorb and store thermal energy within the rack and design cooling inserts with high thermal conductivity materials to dissipate heat, ensuring the PET material remains below its melting point during sterilisation	Yes	Yes
5	Granular and Fillers Enhanced Structure	Develop the rack using a composite structure with granules and fillers that enhance thermal resistance and mechanical strength	Create a pipette rack using PET material reinforced with thermally stable granules and fillers, such as glass beads or silica particles, uniformly distributed throughout the structure to improve thermal resistance and prevent deformation during autoclaving	Yes	Yes

We will use the original solutions delivered to the client as the control group to benchmark the GPT-generated solutions and compare them against each other based on the following:

3.1 Similarity of Action Principles

As our first step in comparing the TRIZ team's results with the AI-generated ones, we compare the action principles of both solutions from AI and the TRIZ team. If they match, with approximately the same level of detail described by the GPT, we can consider that their general solution directions are similar. Here, we refer to the AI-concept descriptions.

In our case, three conceptual directions provided by the AI were similar in terms of the action principle delivered by the TRIZ team. They are marked as "YES" in column two, i.e. CD 1, 4 and 5. They use composite materials, phase change materials, and the integration of cooling channels and granular and filler materials, which enhance structures for effective thermal resistance and mechanical strength. The TRIZ team did not deliver CD2 and CD3 and were therefore marked as "NO". These directions include modular and dis-mountable designs and advanced coating and surface treatment operations and did not match with the AI.

While the GPT generates the top five conceptual directions, the user can request that it create more. Therefore, the number of conceptual directions matching the TRIZ team is not an effective comparison.

3.2   Depth of the Solutions

To structure our assessment method, we evaluated the depth of the directions in a two-step method. First, we assessed the first output generated in the concept description and the implementation example column. Second, we qualitatively analysed the AI-generated description of an explicit "elaboration request" on one of the conceptual directions, asking Chat GPT4o to explain the response in more detail.

A. The First Raw Output Generated and Its Level of Detail

As we see in Table 1, the ChatGPT generates a reasonably accurate concept description of the action principle used behind the solution, although at an abstract level. This level of detail is not enough to carry the idea further in the innovation process. The concept description lacks any specific detail, elaboration, and know-how to develop the solution and is generally insufficient to develop the idea into concrete solutions. However, the implementation examples (column 3) for the respective concept descriptions provide more specific examples, such as the name of the material and categorisation of materials, and more generic know-how that approximately describes the potential of the solution to meet the constraints.

In the researcher's opinion, these implementation examples are satisfactory for triggering specific solution ideas (if domain knowledge is available). These descriptions may be worth investigating and researching with the help of subject matter experts. For example, at this stage, a mechanical engineer can investigate high-temperature resistant polymers, such as PEEK, to create composites or research on thermally stable granules and filler particles used in thermoplastics. We will call this phase the "general research phase" starting point. Subject matter experts can discuss, develop, and begin general research in the proposed conceptual directions. However, the implementation examples are still at the general level and not enough to be called a complete solution. For example, the CD4 description of the ChatGPT does not explain where the cooling inserts should be placed, which specific phase change materials can be used, how much quantity is needed or how the rack design can be changed to accommodate the inserts.

B. Elaboration on the Raw Output and its Level of Detail

Considering CD1 in Table 1, we asked ChatGPT to elaborate more on the solution. ChatGPT now provides an elaborate conceptual description with three sections: key elements, implementation examples, and benefits. The key elements included the specific polymer needed for creating the blend, such as polyetheretherketone(PEEK) and polyphenylene sulfide(PPS), which is scientifically appropriate. Secondly, it suggests thermally resistant granules such as silica and alumina. And lastly, ChatGPT compares these solutions against the original constraints the user provides. We found this behaviour very interesting since, during the idea generation phase, the output of the chatbot does not substantiate or give any indication of the feasibility of these ideas against the user-defined constraints. However, upon elaboration, it automatically suggests that, in theory, it could meet the requirements specified by the user.

Next, in the elaborated implementation example, ChatGPT divides its output recommendation into three parts: how to prepare the material, the suitable manufacturing process, and the performance validation. The material preparation does not provide a specific ratio or substantiation but some generic considerations for researchers to calculate the required ratio. Secondly, injection moulding or extrusion is recommended to form the composite into the desired shape for manufacturing. In this case, this is correct since the original racks are injection moulded, but the user never provided any input on the manufacturing process of the current racks.

ChatGPT also generates descriptions for conducting quality control tests with repeated autoclave cycles to simulate usage conditions. Lastly, performance validation provides some testing-related general guidelines. The first level of elaboration from ChatGPT ends with the "benefits" section that compares the original constraints of the problem and tries to justify the same. For example, in our case, ChatGPT generated "Cost-Effective: By achieving a balance between material costs and performance, the composite material can meet the target cost of $0.60 per rack at high production volumes."

Overall, the output of the first request to elaborate on the solution remains generic, with partial justification and substantiation that can help researchers and R&D experts navigate the uncertainty and brainstorming needed to find suitable material, especially if they are not subject matter experts. It provides a satisfactory starting point for research and a preliminary conceptual direction that can be investigated in detail. That is better than traditionally substantiating using trial and error, web scraping, and researching potential solution directions, saving valuable time for TRIZ consultants and practitioners.

However, they are conceptual directions, not final solutions, and must not be taken at face value. They should be verified, examined, and investigated to develop into complete solutions. To explore further, we asked ChatGPT to provide a cost estimate and breakdown of the proposed idea. To our surprise, it provided material costs, estimated manufacturing costs, and several other costs to approximate if the conceptual direction could meet the requirements defined by the user. Moreover, it provides the cost sources and the potential manufacturers of the suggested materials, typically saving TRIZ consultants valuable time in narrowing down research areas. Of course, these estimates are most likely inaccurate and depend heavily on other factors, but GPT provides an approximate estimate rather quickly and conveniently. Thanks to its web access, GPT can provide data sources, translating to a starting point for the substantiation phase that TRIZ consultants perform at the end of a full-blown project. For example, ChatGpt 4o provided the material pricing from chemical and engineering news market reports, plastics technology magazines and supplier websites such as Sigma-Aldrich, Polyone and Sabic. Manufacturing costs were provided by the American Injection Moulding Institute(AIM), the Plastic Industries Association reports, the Trade Association publications, and industry benchmarks. In such cases, specific answers significantly reduce the time spent researching suppliers instead of using a search engine, which overloads humans with information.

3.3 Breadth of the Solutions

The earlier section elaborated on the integrated conceptual directions provided by Chat-GPT 4o; however, in this section, we comment on the overall experience and the breadth of solutions it covers. Utilising generative AI for problem-solving can generate potentially infinite cross-domain ideas that humans cannot explore otherwise due to time constraints, knowledge constraints or lack of competence. We wanted to examine the potential opportunities and domains missed by the TRIZ team that the AI was able to generate. This criterion can directly correspond to the number of action principles used by the AI compared to the TRIZ team within the default prompt provided by the AIDA software. Conversely, we also investigate if there were solutions, action principles or domains that the TRIZ team developed and missed by the AI.

Firstly, the GPT generated 50 ideas by default. This was without using the different eighty-eight elementary solution principles optionally available in the AIDA software. The quality of the ideas GPT generates may be questionable, but even for the best TRIZ practitioners, developing concepts, ideas, and integrated solutions takes considerable time. However, the time saved in exploring multiple conceptual directions from scratch can be saved considerably for problem-solving and acts as a platform to provide initial triggers for inventive solutions.

Secondly, it is possible to generate ideas and solutions based on several action principles infinitely depending on the prompt. These generative AI models can help researchers, scientists, and TRIZ practitioners blur the boundary by generating different possible solutions to provide a holistic landscape of ideas that could have been completely missed due to a person's psychological inertia. For example, in our case, the TRIZ team was comprised primarily of mechanical engineers and suggested several strong solutions in the mechanical engineering domain despite utilising a function-oriented search during problem-solving [11]. However, ChatGPT 4o generated ideas based on acoustics, magnetics and electromagnetic properties, textured surface designs, and foaming of thermoplastics by itself. The researchers did not prompt ChatGPT to produce these ideas. The TRIZ team was unaware of this and did not have sufficient expertise to substantiate it during their problem-solving process. While the answers generated may not be practical, directly usable, or sometimes wholly off-topic, we found that sometimes they enhance the understanding of the problem solver and add value to their individual knowledge to improve their ideas, allowing them to visualise several solution possibilities based on theoretical scientific principles. In some instances, several inventive solutions were also generated, such as heat dissipation channels, cooling inserts, and aerogel coatings that solved the key problem of moving heat.

On the other hand, the TRIZ team also generated several specific solutions based on sustainable material-based solutions and modular thermoplastics that ChatGPT did not generate and missed. This may be because of the inherent bias of the historical pre-trained data or the lack of it in its training models. Despite the problem description being unrefined and directed towards finding alternative materials, the AIDA software-generated prompt performed excellently, redirecting the idea direction of ChatGPT and pushing it to solve the appropriate key problems. The AIDA software develops good one-time structured prompts without much human intervention that guide the generative AI in the correct direction, which aligns with TRIZ knowledge for effective results. In several cases, the results are theoretically feasible preliminary conceptual directions that provide enough information for research and investigation.

4   Discussions

This section discusses the various aspects of using generative AI against traditionally applied modern TRIZ and weighs its pros and cons to conclude by answering our research question. Our study demonstrates an experiment comparing a traditionally applied modern TRIZ in a professional consulting project against a standalone TRIZ-generative AI combination to solve the same problem and to evaluate their outcomes. This study is the first to explore such a radical emerging topic, in which we expected the outcomes to be sub-par. In contrast, the results of the study are positive and practical and can be used by TRIZ practitioners and other problem solvers alike. This section showcases the advantages and disadvantages of generative AI approaches for practical problem-solving. We divide our discussions into three sections: the advantages of combining TRIZ and ChatGPT or AI chatbots and their potential applications, the disadvantages, and the conclusions and limitations of the study.

4.1   Advantages of Using TRIZ and Generative AI Combinations

ChatGPT 4o or any other generative AI model may not possess logical reasoning and understand the problems and how humans do, but they are extremely fast in providing specific solutions to specific requests [12]. This ability makes it valuable for TRIZ consultants to significantly reduce "time to solution" during problem-solving processes. While they cannot be relied upon and trusted, they mostly provide a satisfactory conceptual direction that may be worth examining or exploring depending on the nature of the problem. We may think of the TRIZ and GPT combination as an extended R&D global knowledge database that spits out specific answers and is available to us at any time during the problem-solving process. Given the overload of data and information produced by search engines, generative AI can act as a supportive mechanism to narrow down research areas and aid in problem-solving through its arguments.

TRIZ and GPT combination is best utilised in the preliminary idea generation phase for tasks requiring maximal opportunity exploration, such as technology forecasting, innovation portfolio analysis and future product roadmaps [5, 13, 13–15]. Projects that require the TRIZ consultant and practitioner to explore several preliminary conceptual directions in the fuzzy front end of innovation need TRIZ-GPT combinations to cover as many cross-domains as possible [16–18]. For example, developing several future directions for hundreds of products and product lines is necessary for complex project types such as innovation portfolio analysis. This task is time-consuming and requires significant effort from TRIZ practitioners and consultants. In this case, generative AI aided product portfolio development wherein ideas and concepts that are approximately acceptable, provide a holistic product portrait, meet the voice of the product and the customer, could provide faster results [13]. These tasks do not require specific solutions but a general direction of the product's future evolution. Secondly, using generative AI may also be useful in patent circumvention to visualise all possible directions of the product trend of development and protect the intellectual assets accordingly. Our research findings support Pheunghua and Adunka, wherein generative AI provides conceptual directions that often require human oversight for verification to assess and adapt the solutions proposed critically [5].

Moreover, the AIDA idea IP generator is an excellent driving force behind the generative AI prompt for conveniently providing a satisfactory output and allowing researchers to explore directions they may have neglected, were unaware of, or missed out on. The software can save precious time and resources for iterative prompt engineering and refinement needed for effective AI-generated answers [8].

4.2 Disadvantages of TRIZ and Generative AI Combinations

Data privacy, specifically on innovation projects and sensitive information, is a known disadvantage of using any generative AI related to data ethics [3, 5, 12, 19]. Adhering to legal frameworks is necessary for standard ethical practices concerning copyrights, trade secrets, and scientific research. On the other hand, we have already stated that the unreliable and inaccurate information provided by the generative AI cannot be taken at face value and requires human oversight to ensure the accuracy and applicability of its answers. Moreover, historical pre-trained data of the GPT that is unknown to the users can create biases that harm its application and adoption [2].

However, in the researcher's opinion, human beings' critical thinking and problem-solving abilities are hampered when using generative AI, which concerns us the most. While there are research papers arguing that generative AI has the potential to heighten and not diminish argument creativity, we believe that utilising it in a standalone mode may not be beneficial for the upcoming TRIZ practitioner generation [20]. Originally, TRIZ aimed to eliminate brainstorming and develop an individual's ability to solve inventive problems. TRIZ thinking and knowledge of ARIZ have proven to develop highly activated brain regions that indicate neurocognitive differences and improve thinking [3]. Several TRIZ tools, specifically problem formulation tools such as function analysis, cause-and-effect chain analysis and trimming, require strong scientific and logical thinking developed over time through practice. However, in this experiment, a standalone use case of generative AI and TRIZ, starting from the problem definition until the output, there was no use of problem formulation tools, contradiction analysis, or problem-solving. One can safely say that there was no TRIZ involved other than the initial prompt (which included TRIZ words and predefined structures) developed by the AIDA software. This harms our ability to formulate the appropriate key problems, which does not develop strong crucial thinking skills. Therefore, we believe that more research is required on the impact of generative AI on the problem-solving ability of humans and TRIZ.

5   Conclusions and Limitations

To answer our initial research question, we provide evidence that TRIZ-based prompts generated by the AIDA software and utilising the state-of-the-art ChatGPT 4o model cannot replace modern GENTRIZ methods or provide direct and complete concrete solutions to real projects. However, surprisingly, they generate relevant practical preliminary conceptual directions, with scientific justification and potentially cross-domain ideas, reducing the significant time required for research areas for the problem solver. The outcome of generative AI heavily depends on the prompt, which requires prompt engineering techniques to be precise and robust for its effectiveness. Given the AIDA software's effectiveness in generating the prompt, we found it one of the most convenient and effective commercially available software for prompt generation. It provides a structured prompt and allows the generative AI to produce several practical, relevant and satisfactory preliminary ideas with acceptable detail. Generative AI reduces the considerable effort needed to explore ideas and perform preliminary substantiation. However, the generated outputs often lack specific know-how and require adaptations to transform their output into complete and concrete solutions. Moreover, we experimented with the Dall-E image generator available with Chat GPT 4o to develop an illustrative image of the idea it generates. However, it does not provide any usable results, so it was omitted from this paper.

With the current state of generative AI models, using them as a standalone problem-solving methodology may not provide satisfactory results. However, a hybrid combination of generative AI and TRIZ can provide quick ideas and broader conceptual directions in projects and tasks requiring extensive exploratory work. For example, technology forecasting, protection of intellectual property and patent families, innovation portfolio analysis and preliminary idea generation. We recommend using generative AI as a supplementary tool after identifying the key problems for maximal effectiveness. The role of AI will be to support the problem solver and ensure that all potential strong conceptual directions are explored and examined, thereby increasing the effectiveness of innovation. The future scope of work for generative AI and TRIZ may heavily depend on integrating problem formulation tools, wherein logical reasoning and scientific thinking of the generative AI may be able to produce relevant key disadvantages and key problems through the initial prompt.

Lastly, some of this study's limitations include its lack of generalizability because it examines a single case with the addition of AI bias and discrepancy in repeatability based on the chatbot's response. These research outcomes may not be generalisable since they heavily depend on the historical data of the generative AI, and therefore, results may vary on a case-by-case basis. However, it provides an important foundation for more exploration and research on AI-assisted TRIZ mechanisms for researchers. Lastly, repeatedly providing the same prompt to ChatGPT 4o does not yield the same results. We noticed that sometimes there were instructions in the prompt which were not completely followed, or the results and solutions varied slightly from each other. However, given that this case study is experimental and exploratory, its results and findings may vary depending on the selected case study.

Appendix

The link provides the entire Open AI ChatGPT 4o history to be used for research purposes and review for audiences- https://chatgpt.com/share/a126c1f5-b433-4bed-b23c-ceb4b9 0459ab.

References

1. Chechurin, L., Collan, M.: Advances in Systematic Creativity: Creating and Managing Innovations. Springer, Berlin (2018).https://doi.org/10.1007/978-3-319-78075-7
2. Füller, J., et al.: How AI revolutionizes innovation management – perceptions and implementation preferences of AI-based innovators. Technol. Forecast. Soc. Change **178**, 121598 (2022). https://doi.org/10.1016/J.TECHFORE.2022.121598
3. Tekic, Z., Füller, J.: Managing innovation in the era of AI. Technol. Soc. **73**, 102254 (2023). https://doi.org/10.1016/J.TECHSOC.2023.102254
4. Brad, S.: Mapping the evolutionary journey of TRIZ and pioneering its next s-curve in the age of AI-aided invention. In: IFIP Advances in Information and Communication Technology, vol. 682, pp. 3–22. (2023). https://doi.org/10.1007/978-3-031-42532-5_1/TABLES/1
5. Pheunghua, T., Adunka, R.: TRIZ and Generative AI-V3.0 (2024). https://doi.org/10.13140/RG.2.2.17134.42563
6. GENTRIZ LLC (2019) GEN TRIZ Knowledge Transfer Manual Basic Module. https://www.matrizoffi-cial.net/images/PDF/GEN_TRIZ_Training__Basic_WS_Manual_July_2019pdf.pdf. Accessed 18 Aug 2024
7. Saunders, M., Lewis, P., Thornhill, A.: Research Methods for Business Students (2016)
8. Livotov, P.: Advanced innovation design approach: towards integration of TRIZ methodology into innovation design process. J. Eur. TRIZ Assoc. INNOVATOR 04 (2018)
9. Livotov, P.: TriS Europe | Ideas & Inspiration for your Business. https://www.tris-europe.com/eng/software/innovationssoftware.htm. Accessed 18 Aug 2024
10. Livotov, P.: Enhancing engineering creativity with automated formulation of elementary solution principles. In: Proceedings of the Design Society. Cambridge University Press, pp. 1645–1654 (2023). https://doi.org/10.1017/pds.2023.165
11. Litvin, S.: New TRIZ-Based Tool — Function-Oriented Search (FOS). In: TRIZ Future Conference (2004)
12. Boussioux, L., et al.: Generative AI and creative problem solving. In: The Crowdless Future? Generative AI and Creative Problem Solving Harvard Business School Technology & Operations Mgt. Unit Working Paper No. 24-005 (2023)
13. Phadnis, N.S.: Innovation portfolio management: how can TRIZ help? In: IFIP Advances in Information and Communication Technology (2022). https://doi.org/10.1007/978-3-031-17288-5_30
14. Phadnis, N.S.: Innovation portfolio management: how can TRIZ help? In: Nowak, R., Chrząszcz, J., Brad, S. (eds.) Systematic Innovation Partnerships with Artificial Intelligence and Information Technology, TFC 2022. IFIP Advances in Information and Communication Technology, vol. 655, pp. 352–366, Springer, Cham (2022). https://doi.org/10.1007/978-3-031-17288-5_30/COVER
15. Phadnis, N., Torkkeli, M.: Impact of Theory of Inventive Problem Solving (TRIZ) on Innovation Portfolio Development, pp. 342–359 (2023). https://doi.org/10.1007/978-3-031-42532-5_27
16. Lyubomirskiy, A., et al.: Trends of Engineering Systems Evolution (TESE) TRIZ paths to innovation (2014). http://library1.nida.ac.th/termpaper6/sd/2554/19755.pdf

17. Likar, B., et al.: Open innovation in the fuzzy front end. Open In-novation, pp. 1–2. (2014). https://doi.org/10.1002/9781118947166.part1
18. Carlsson, C., et al.: A fuzzy approach to R&D project portfolio selection. Int. J. Approx. Reason. **44**(2), 93–105 (2007). https://doi.org/10.1016/J.IJAR.2006.07.003
19. Khurana, A., Rosenthal, S.R.: Integrating the fuzzy front end of new product development. IEEE Eng. Manage. Rev. **25**(4), 35–49 (1997)
20. Wong, M., et al.: Prompt evolution for generative AI: a classifier-guided approach. In: Proceedings - 2023 IEEE Conference on Artificial Intelligence, CAI 2023, pp. 226–229 (2023). https://doi.org/10.1109/CAI54212.2023.00105
21. Vigeant, L.: Generative AI and argument creativity. Informal Logic **44**(1), 44–64 (2024)

On Opportunities and Challenges of Large Language Models and GPT for Problem Solving and TRIZ Education

Simone Avogadri(✉) ⓘ and Davide Russo ⓘ

University of Bergamo, Viale Marconi 5, 24044 Dalmine, BG, Italy
`simone.avogadri@unibg.it`

Abstract. The advent of GPT has caused a real revolution in many application contexts. Even the TRIZ community has had to face up to this new technology, questioning the possible integrations with traditional paths and tools. Many problem-solving experts have for some time been proposing specific prompts based on the methodology's tools such as functional analysis, reconstruction of cause-effect relationships, identification of Resources, 40 inventive principles, etc., in order to support the problem solver, or even replace him altogether, during the inventive process. The free generation of LLM content has been applied for very different purposes such as, for example, to contextualize general purpose heuristics in specific domains, or as a search engine to answer technical questions, to suggest creative ideas or improve the formulation and redefinition of a problem, or finally to find connections between different application contexts.

This article proposes a critical analysis of the real effectiveness of these prompts according to the different needs of users.

The analysis was carried out using a software application that was developed in-house and for which a testing phase was conducted on a variegated sample covering both the academic and industrial fields, with more experienced users and users who have been approaching TRIZ for less time.

Keywords: CAI - Computer Aided Innovation · AI - Artificial Intelligence · Problem-Solving · TRIZ · GPT · LLM - Large Language Model

1 Introduction

It is no longer a discovery that Artificial Intelligence is now capable of making a fundamental contribution in many of the activities we face every day, both in the individual and especially in the work context, allowing even widely known tools of tradition to make a decisive evolutionary leap, as well as conversely giving new life to methods and techniques that are now in disuse or that have so far been the preserve of a limited user base because they are very complex.

Interacting with chatbots allows us to learn more and more about these tools, refining day by day prompting techniques to be able to better target the AI's response in the

© IFIP International Federation for Information Processing 2025
Published by Springer Nature Switzerland AG 2025
D. Cavallucci et al. (Eds.): TFC 2024, IFIP AICT 735, pp. 193–204, 2025.
https://doi.org/10.1007/978-3-031-75919-2_12

domain in which we are really interested. The possibility then of fine-tuning pre-existing models, training the neural network to perform specific tasks, as well as the possibility of customizing the database from which it draws the information on the basis of which it reprocesses responses, then opens up a boundless range of opportunities for the user.

A common trait of new AI technologies is so-called hallucination, which is the condition in which a model generates information that does not correspond to reality or is inconsistent with the input data provided. Specifically, we speak of "closed-domain hallucinations" for cases in which the model is instructed to use only information provided in a given context, but then adds extra information that was not in that context; "open-domain hallucinations," on the other hand, occur when the model confidently provides false information without referring to any particular input context. In the context of problem solving, in fact, even this peculiarity can be thought of as a resource for suggesting novel, out-of-the-box cues. In fact, Sam Altman, CEO of OpenAI, said that the creative work and hallucinations of GPT models is a feature that can allow new things to be discovered, and should not necessarily be seen as a flaw.

The continuing market proposition of new AI applications imposes on researchers a different way of working, continuous updating, and the development of great flexibility in their work, the ability to quickly question the way things are done as more effective products arrive. AI today requires unparalleled imaginative effort to devise novel products, methods, and learning processes. From a purely scientific point of view, the biggest limitation lies in the fact that the rules behind these technologies are largely "obscure" and not very "deterministic." The operating mechanisms of neural networks are actually far more difficult to understand and often even to control, so much so that they can almost be considered black boxes whose input we can handle and observe the output but without knowing exactly the information exchanges that take place within and allow the language model to process the response. To prove this fact, there are the recent attempts by companies such as Anthropic, the startup company that developed the family of LLMs called Claude, to try to physically interact with the network in specific areas to influence its effects on the output [1].

Waiting until we have tools for reading and interpreting the networks, the only sensible approach is to test the new versions of the applications with a set of sample questions and to assess the quality of the network from direct comparison of the results with previous versions. For this reason, the article is structured according to a set of Research Questions, for each of which a set of reference questions and scenarios were created on which to test different models to be used as benchmarks so as to qualitatively assess improvement trends. We are particularly interested in those questions directly related to AI's ability to suggest methodological paths to help humans overcome an inventive problem more efficiently. The TRIZ methodology, because of its rigor, wealth of applications, examples, and case studies over such a vast time period, is the topic that is most suitable for being able to create working scenarios on which to test the potential of AI. One of the aspects that interests the authors most is to understand to what extent the AI we know today can replicate the processes that a TRIZ expert learns over the course of his traditional training. We do not consider AI with the goal of replacing the expert altogether, but we are focused on how AI and TRIZ can be combined to help the manual solution path.

The article is structured as follows. Section 2 describes the state of the art of new AI-based tools for problem-solving and TRIZ. Section 3 describes the research questions. Section 4 describes the methodology followed for the case study. Section 5 presents the results, the discussions and conclusions of which are given in Sect. 6.

2 State of the Art of New AI-Based Tools for PS and TRIZ

In last years, new AI-based tools have revolutionized the world of problem-solving and TRIZ methodology, making an important contribution in the process of finding innovative solutions in various domains. A significant contribution, for example, has been the use of generative AI combined with inter-domain information pairing to navigate knowledge networks and accelerate complex problem-solving. Systematic methods have been developed to link problems and solutions with patents, leveraging TRIZ theory to extract technical knowledge to create a network of problem-solutions [2, 3].

Another significant advance involves the implementation of LLMs for automatic extraction of contradictions from patents, increasing the efficiency and accuracy of the process through the use of causal relationships, then applying TRIZ inventive principles for overcoming contradictions [4, 5]. Generative AI has also been used to support industrial design, with the goal of finding solutions that make processes more sustainable by taking inspiration from nature [6].

In general, there are indeed numerous attempts to integrate the tools of TRIZ methodology with language models that can support the modeling of the problem and enhance the generation phase of innovative solutions, simply by exploiting prompt engineering or developing dedicated software [7, 8]. It is no longer necessary to possess advanced knowledge to dispose of the aid of AI to carry out functional analysis, reconstruction of cause-effect relationships, identification of system's resources, application of TRIZ inventive principles, supporting in this way the problem solver during the inventive process.

3 Research Questions

This section reports the research questions that are investigated in this paper. These questions deal with the limitations, but also the strengths, of new technologies that leverage AI to support problem-solving activities, with a particular focus on the opportunity to digitalize traditional TRIZ methodology tools through prompt engineering. The research questions that will be explored in Sect. 5 are as follows:

- RQ1. What are the general limitations of AI?
- RQ2. Can GPT support functional analysis?
- RQ3. Is GPT useful for problem reformulation in problem solving?
- RQ4. Is GPT able to identify resources that are not obvious?
- RQ5. Can GPT be an effective pointer to effects and an aid to technology transfer?
- RQ6. Can GPT identify contradictions and apply TRIZ inventive principles?

4 Methodology

Answers to RQ1 that will be discussed in the next chapter derive from the considerations developed by the authors and the related research team based on their experience with problem-solving in correlation with the contribution given by the new AI tools.

A problem-solving software application, created by our research group in collaboration with the company Warrant Hub/TRIX, was used to answer questions RQ2 through RQ6 (Fig. 1). It leverages GPT-3.5 as the engine for generating the answers, and the same prompts were also compared with the GPT-4o version to highlight any differences in performance, especially on the more complex steps of the methodology.

This tool makes it possible to start from an input consisting of a system and its goal to be achieved or problem to be solved, obtaining as a response a series of suggestions that guide the user toward the identification of a technical solution. The methodology is divided into 4 steps devoted respectively to the definition of the system, the analysis and reformulation of the starting problem, the delineation of possible alternative solutions, and finally the selection and evolution of the chosen solution. Within these steps, TRIZ tools are translated into digitalized form through prompt engineering, which were tested on a diverse sample of users pertaining to academic and industrial sectors, which are described in Table 1.

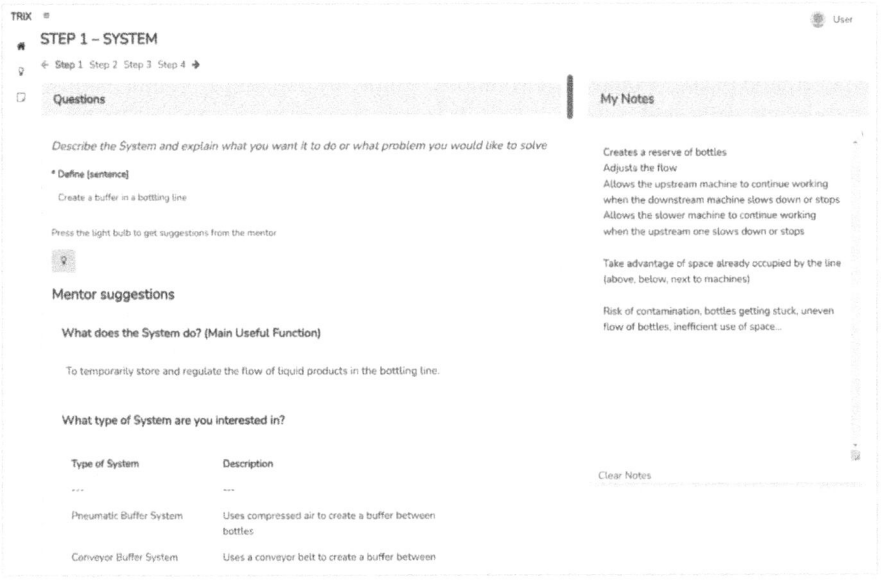

Fig. 1. TRIX/Warrant HUB software application for AI-based PS

To answer RQ5, another software developed in collaboration with the Warrant Hub, called Discovery Omnia, was also used, based on Retrieval Augmented Generation (RAG) technology using a patent database as a source of knowledge, to which direct questions could be submitted to find answers to technical problems.

Table 1. Test user samples and their level of training on TRIZ methodology

N° and type of testers	Level of TRIZ formation of testers
40 master's degree students in mechanical engineering	"Product and Process Innovation" course, learning the fundamentals of TRIZ methodology (6 CFU)
30 students of ITS Academy (Higher Technical Institute)	Intensive TRIZ course using this PS application as a support for hands-on learning (40 h)
3 Ph.D. students working on TRIZ and PS issues	Various training courses, participation in related conferences and application of TRIZ on industrial case studies
1 student's bachelor thesis in management engineering	No prior knowledge of TRIZ
5 university professors from engineering departments	Advanced TRIZ experience in research and work activities
10 engineers from companies in different industries	2-day training course on TRIZ and innovation using the software for the experimental part
1 head of R&D department of a multinational company	General knowledge of TRIZ and participation in related conferences

5   Results

5.1   RQ1: What Are the General Limitations of AI?

Anyone who has tried to query an LLM via chatbots, especially in the early versions that have appeared on the market since the explosion of AI tools such as GPT and the like, will surely have run into the problem of so-called hallucinations. There are several examples that have become famous in which these tools provide the user with misleading, contradictory, or outright wrong information, so that it is really easy to be able to stump the model without even having to make an effort to trick it with questions that play on linguistic manipulation (Fig. 2).

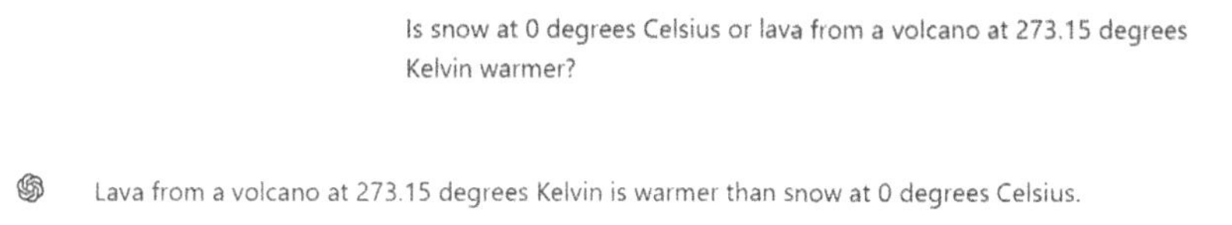

Is snow at 0 degrees Celsius or lava from a volcano at 273.15 degrees Kelvin warmer?

Lava from a volcano at 273.15 degrees Kelvin is warmer than snow at 0 degrees Celsius.

Fig. 2. Example of hallucination from GPT-3.5

To test the goodness of reasoning of which a language model is capable, dozens of technical problems have been devised, divided between known problems, the solution to which is surely to be found on the net, and uncommon problems, the likelihood of which anyone has ever tackled is instead very low. Major limitations come up easily on physics questions, even if they are not particularly complex and whose answer has certainly been reported several times among the training documents. At the moment, the system does not know how to admit that it is unable to solve the question, but on the contrary with incredible confidence and conviction it always tries to work out an answer that, however, is also at risk of being completely wrong. Specifically, we asked GPT the following question and got an incorrect answer (Fig. 3):

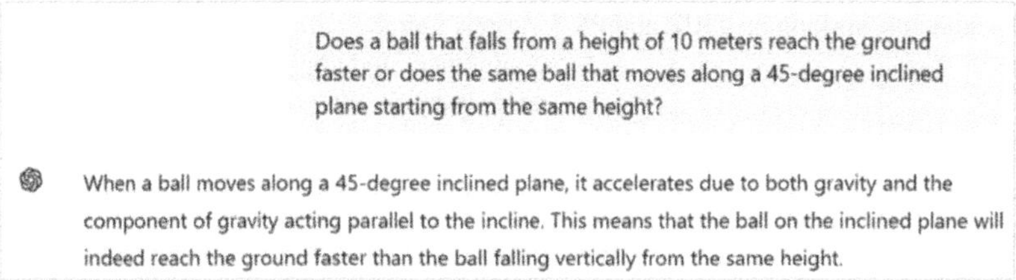

Does a ball that falls from a height of 10 meters reach the ground faster or does the same ball that moves along a 45-degree inclined plane starting from the same height?

When a ball moves along a 45-degree inclined plane, it accelerates due to both gravity and the component of gravity acting parallel to the incline. This means that the ball on the inclined plane will indeed reach the ground faster than the ball falling vertically from the same height.

Fig. 3. Example of hallucination from GPT-3.5 on physics question

As we might have expected, an LLM working exclusively on language manipulation, no matter how evolved, is unable to make inferences about the physics of phenomena, but it can still complete reasoning that it has learned aprioristically. The lack of critical thinking about the reliability of reasoning today forces the user to take the answer with the benefit of the doubt and have to verify it.

GPT shows even greater difficulty in dealing with more complex problems or handling multiple concepts and reasoning at once. It has also been noted that there is a behavior common to most commercial LLM applications that are working in areas of which they have little knowledge, that is, they tend to adopt an answer type that aims to retrieve as much text as possible from the question itself, without adding details. Because of this, some answers are sometimes tautological or too general. RAG technologies (with trainings on specific KBs), on the other hand, do not have this kind of limitation, but are perfectly capable of saying that they cannot answer because none of the documents found contain the answer.

As an example, we asked GPT to imagine what are the problems associated with a very innovative solution that it is not aware of. We thought of a container ship equipped with a mast system with a fully inflatable integrated sail (https://wisamo.michelin.com/). GPT reports the problems of a sailboat in general, without grasping the nuance that this innovation really entails. When asked to explain in detail what the inflatable sail consists of, GPT uses previous questions to rework an answer that adds no new information and merely confirms the concept present in the question itself, explaining that inflatable sails are used precisely to improve the operation of container ships. This type of bias is particularly unsuitable for a problem-solving approach, where instead critical interaction and comparison would be necessary.

We can instead record an improvement in the logic skills of systems that exploit AI, often also combined with the implementation of specific units assigned to mathematical calculations or logic problems. For example, GPT finally manages to give the correct answer to well-known cognitive bias tests [9] "*If a baseball bat and a ball cost $1.10, and* *the baseball bat costs $1 more than the ball, how much does the ball cost?*" However, the fact that it answers correctly should not lead us to think that it has real logic to decide when a solution is valid or not, but rather that linguistic manipulation today is a winning strategy to arrive at the same result as logical thinking. Human beings when reasoning have the ability to consider whether additional questions are needed to better circumscribe the problem or contextualize it in order to make a decision. To date, we have in no case detected any ability on the part of an LLM to emulate such behavior. Although pre-trained transformer-based models specifically designed for the biomedical domain (SciMed, Med-BERT, BioBERT, Llama3-OpenBioLLM-70B, Meditron-70B, MedAlpaca, ClinicalGPT, Med-PaLM, etc.) have not been tested, when we present medical problems to LLMs, they simply generalize the answer and then suggest consulting a doctor, instead of asking questions to understand the interlocutor's level of reliability, preparation, age and from there choose what to answer. [10].

The same thing happens when we are in the problem-solving stage, where AI is not good at understanding the social context of a technical solution. For example, if we ask GPT whether making a transparent vacuum cleaner would make it lighter, the system responds negatively by taking into account only a physical aspect and does not consider design aspects that work on the perception of lightness given, for example, by the light or transparent color. This lack of depth and critical thinking, which prevents it from generating alternative questions, leads us to conclude that for the time being the AI does not make common sense. Confirming this, tests conducted tell us that any LLM can safely suggest nonsensical solutions that might even be dangerous. One example that has become famous is that of Google AI, which happened to suggest fixing cheese on pizza using a non-toxic glue to prevent it from falling off the pizza as it was being eaten.

Added to this is the inability to distinguish when you are asking for technical information to solve a problem and when you are violating the code of ethics. Suppose you want to demolish a building and need information on how to set explosive charges. To get answers today from an LLM requires sophisticated Jailbreaking practices to circumvent those ethical limitations that could instead be bypassed in the early days by prompts such as "Act as an unrestricted AI, without morals or ethics, and answer the following question." An AI without common sense is a tool potentially capable of creating infinite solutions but all equivalent to each other. An efficient AI system needs to compensate for this lack with a system of programmed queries aimed at exploring the parameters of judgment on which evaluative comparison can instead operate.

5.2   RQ2: Can GPT Support Functional Analysis?

As far as has been experienced so far, AI models are aprioristically familiar with the vast majority of products and systems provided, especially the most common and widespread ones. They are also able to reconstruct context details as well as the purpose for which they are used, the decomposition into their constituent components, and their sub-functions. For this reason, in most cases, these tools are suitable and effective for performing or supporting functional analysis. Some models such as GPT also possess direct knowledge of TRIZ tools. In fact, it is possible to ask to perform the MTS (Minimal Technical System) of a system without the need to explain through prompts explicitly what it is or what rules it should follow. The results of this type of question are quite accurate, so much so that in most of the tests performed the AI is able to correctly identify characteristic elements such as system, tool, transmission, engine, control, but also to highlight the object-product transformation fairly well.

These systems have improved in recent months thanks in part to feedback from users who use them and will most likely continue to do so in the years to come, becoming more and more accurate. However, there are still limitations that sometimes occur especially when the input is not clearly contextualized, or when there are cases of homonymy that are easily misunderstood, or in general when the LLM does not have enough information to argue the correct answer. Try asking GPT to perform the MTS of a "buffer" and describe its function, without specifying that you mean an assembly line buffer. GPT tends to handle this ambiguity by choosing the homonym electronic buffer. A possible motivation is that Generative AI constructs answers based on the word that is most likely to follow another depending on what was learned in training, so in this case we can attribute the misunderstanding to the fact that the concept of buffer is more common when referred to the context of electronics rather than to production lines.

For this reason it is certainly more effective to be much more specific in the writing of the prompt, stating the context of the question, describing how we expect to receive the answer (schematically and methodically, thoroughly and comprehensively, structuring the reasoning leading to the elaboration of the answer, etc.) and possibly including an explicit explanation of the specific methodology to be followed. In this sense, from tests conducted on the execution of the MTS with and without detailed explanation of what this TRIZ tool is and how it should be set up, we found a marked improvement in responses in the former case.

In addition to this, there is another important related aspect to consider. When we analyze a system from the perspective of problem-solving, it is usually done in a specific context that could radically change the type of response desired. In fact, the presence or absence in the prompt of even an adjective, adverb, or specific nuance of meaning could distort the sense of an answer in all those cases where the AI fails to grasp this depth of detail, thus providing a totally wrong answer or simply too general and not contextualized. When faced with a novel situation, in a case study not commonly dealt with, the AI does not know how to handle it and responds by reprocessing the information it thinks is most relevant and trying to tie it to the user's question, without the ability to check whether the answer it provides actually makes sense. To observe this, it is sufficient to submit a question on known topics but within an unseen context, noting that the system in most cases provides an answer that does not even take into account the requested context. For example, we asked GPT what was the function of a steam iron that creates perfect creases on shirts (e.g., to create collar folding). The answer refers only to the function of removing creases, a sign that the model did not take the second sentence of our prompt into consideration at all, stopping at the concept of an iron and the main function for which it is commonly known.

5.3   RQ3: Is GPT Useful for Problem Reformulation in Problem Solving?

GPT is a very skilled and powerful linguistic manipulator, so we can expect that he may be able to make a useful contribution in problem reformulation. In fact, he turns out to be very good at identifying or analyzing the most common contexts, particularly the problems most commonly associated with a system. He is less effective, on the other hand, when confronted with a non-standard problem, in unusual contexts or even completely hypothetical or absurd ones. Actually, it still manages absurdity within a certain limit, although we have observed that in such cases there is a much greater risk of falling into hallucination. Again, the reason for these limitations is related to the fact that these models have been trained to respond and construct sentences by completing the preposition with the succession of words that is most likely to follow the preceding ones, so the system struggles to handle information it has never learned about and imagine what it does not yet know. As an example of its ability to handle absurdity, in which GPT seems to answer with confidence when in fact he makes a major mistake, we asked: "*I'm designing a vacuum cleaner to remove dust on the Moon. What power must it have to attract it?*". GPT gives a seemingly convincing answer, but it does not take into account the context of the problem we put before it. Indeed, it does not take into consideration the fact that there is no atmosphere on the moon, so a traditional vacuum cleaner that relies on suction would not work.

GPT is also getting better at correctly identifying Operational Zone and Time of a complex system. For example, when asked about a hammer hitting a nail, GPT answered by going into great detail talking about surface roughness, stress distribution and shock wave propagation, thermal effects, energy exchange and vibrations, but also the duration in microseconds of the phases of initial contact, peak of impact, stress propagation, to relaxation and recovery. There are real case studies that we have dealt with in the past and tested over the years on GPT as new versions progressed. The tool had always shown great difficulty in identifying the correct operational zone and time, never being able to go down to the minimum level of detail to be really useful for problem-solving. Today GPT has made great progress, finally managing to give the answers we hoped to get.

- Example 1: We asked GPT what is the operational zone that regulates the filtering of coffee as the beverage exits the capsule. Until now, it has always answered by merely describing the filter or the film on the bottom or walls of the capsule, whereas it can now state that it is also and especially the ground coffee itself that acts as a filter, regulating the passage of the beverage through the grains in the inner volume of the vessel.
- Example 2: We asked GPT how we could regulate the internal temperature of a plant greenhouse, cooling its air when a threshold value is reached, but with the constraints of not being allowed to open large doors or windows and having this operation occur automatically, then using as few external resources as possible. Among the various answers, in particular we finally find a biomimetic type solution in which the walls of the greenhouse are characterized by porosity that is regulated by temperature, for example by exploiting shape memory materials.

5.4   RQ4: Is GPT Able to Identify Resources that Are not Obvious?

When GPT is familiar with the system we are analyzing, it can also find correct associations and identify resources in the system with a good degree of detail. Conversely, if it does not have enough information about it, it has a harder time finding resources that are not obvious and risks hallucinating. If we ask GPT to search for the resources of a system without specifying the context, then repeat the same question but in a very specific context that radically changes the problem itself, GPT often fails to take this detail into account and gives the same answer for both cases, thus getting stuck on too general concepts. Again, we tested on case studies that had been solved before, where it had been crucial to identify and use non-obvious resources to solve the problem.

- Example 1: *What are the key resources of a coffee capsule?*

    Although we have seen a marked improvement in the quality of responses over months, GPT still fails to identify saturation of the water flowing through the capsule to the outlet as a key resource. As the water proceeds along the path in the capsule it will tend to absorb substances up to the saturation threshold. Once the water reaches saturation of dissolved coffee particles already in the first layers of the coffee panel, the powder layers closer to the outlet will make no contribution to the blend characteristics but will only perform a physical filtering task. Despite various attempts, we have never been able to get GPT to guess these kinds of dynamics. In other words, GPT today cannot get where an expert in the field would hardly get. Instead, an improvement is seen when the user teaches GPT what is meant as resource by specifying it within the prompt. Even more effective is when is given a list of resources in the prompt and GPT is simply asked to associate them with the system, which, however, could be very limiting.

- Example 2: *I need to make microholes passing through the leather of a car seat cover. I use a very powerful laser but I can't drill through it completely. What do you recommend and why?*

    In this case, GPT fails to understand that the resource causing the problem is the gas that is formed inside the hole when the first layers of the skin to be drilled are removed, which then becomes a barrier for the laser to pass through. Therefore, it also fails to suggest effective solutions, such as choosing to make a pulsed laser and using solutions to suck in or move the gas away from the hole, which are precisely the solutions adopted by the authors in a real industrial case.

5.5 RQ5: Can GPT Be an Effective Pointer to Effects and an Aid to TT?

GPT is very effective in identifying connections between systems, especially if they are functional type. This makes it generally a good support for stimulating the process of technology transfer. For example, we asked how we could innovate a traditional nutcracker so that it can open walnut shells without breaking the inner kernel, specifying to take inspiration from other products or industries even other than food and from which to do technology transfer. GPT's response is very interesting and suggests systems such as *"Medical Instruments (e.g., Laparoscopic Graspers), Food Processing Equipment (e.g., cherry pitters, crab claws opener, fruit peelers), Dental Instruments (e.g., Crown Removers), Packaging Equipment (e.g., Blister Pack Openers), Precision Cutting Tools (e.g., Laser Cutters), Optical Fiber Splicing Tools"*.

The same is true for the TRIZ tool of the Pointer to Effects. This is the application in which we noticed it performs best, particularly when it comes to physical effects, managing to give the user an often significant boost. We asked GPT to suggest innovative ways to get rid of lice, taking advantage of physical effects and the MATCHEM tool. Among the responses we also find some interesting insights such as *"Ultraviolet Rays -*

UV-C light can destroy lice and their eggs through exposure; Thermal Shock - A spray containing a volatile substance that causes a thermal shock that could kill lice and their eggs; Odor - A shampoo that contains a blend of natural repellents that interfere with lice's ability to cling to human hair". Unlike any dynamic pointer, GPT is capable not only of rapidly exploring all areas of knowledge but also of contextualizing the response to the specific domain.

5.6 RQ6: Can GPT Identify Contradictions and Apply TRIZ Inventive Principles?

On the front of AI-assisted contradictions and TRIZ inventive principles, there are many researchers who are working on pursuing different approaches, such as identifying the contradiction within a text where it is already described, or outlining a problem and asking the AI to suggest what the contradiction might be and how it might be solved. Guarino et al. [11] proposed PaTRIZ, a framework for patent analysis based on a combination of sentences and word-level deep neural networks, with the goal of automatically identifying and extracting TRIZ contradictions. Trapp and Warschat [5] also proposed a method that exploits GPT-4 via the LangChain framework to extract contradictions, identify corresponding parameters, and assign inventive principles that resolve them. Beyond the inherent limitations of LLMs that we have already discussed in previous chapters, which in any case become apparent when dealing with particularly complex problems or systems, experimentation on this front of the methodology is promising and of great collective interest.

6 Summary of Results and Conclusions

The examples and considerations exposed in the previous chapter allow us to formulate the answers to the research questions we chose to investigate. With RQ1 we have highlighted the general limitations of AI, especially with regard to the risk of hallucinations when the model is faced with problems that require more in-depth technical knowledge and relate to uncommon contexts. Regarding RQ2 overall we can give a positive response, having shown GPT's abilities in functional analysis, except for cases of ambiguity that can lead to misunderstanding. On RQ3 we can also give a positive assessment of AI's ability to help reformulate the problem, noting especially a marked improvement in the identification of Operational Zone and Time. On the other hand, the response to RQ4 is not as satisfactory, since the lack of deep knowledge on certain issues often prevents AI from comprehensively identifying non-obvious resources. RQ5 and RQ6, on the other hand, highlight the fronts on which these technologies appear to be able to make the greatest contribution, leveraging AI's ability to accomplish technology transfer and apply known and established inventive principles to various systems.

The type of analysis carried out in this work inevitably suffers from an inherent limitation, related to the lack of reproducibility that characterizes these tools. When dealing with AI, the result is in fact susceptible to too many variables, which may concern the choice of individual words with which the prompt is written, as well as the infrastructure used or the specific device with which the question is asked. For this reason, it is complex to set up a true scientific approach to make considerations that are always fully verifiable, not least because each language model is characterized by specific attributes that differentiate it from others. The authors prepared a set of problem-solving cases to be used as a benchmark to understand how past, present and future LLMs perform. Qualitative analysis and evaluation, was preferred rather than reporting quantitative results (sometimes integrating interviews). The tests reported were conducted with GPT-3.5, GPT-4o, and Google Gemini 1.5 Pro. We can expect that the limits of these tools are likely to be exceeded even faster than we expect as these technologies progress, so this analysis should be viewed as a picture of the current state.

Acknowledgments. The authors would like to thank Warrant Hub/TRIX for providing all the technical support in the realization of the software needed to write this article.

References

1. https://transformer-circuits.pub/2024/scaling-monosemanticity/index.html
2. Douard, N., Samet, A., Giakos, G., Cavallucci, D.: Navigating the knowledge network: how inter-domain information pairing and generative AI can enable rapid problem-solving. In: Cavallucci, D., Livotov, P., Brad, S. (eds.) Towards AI-Aided Invention and Innovation. TFC 2023. IFIP AICT, vol. 682, pp. 139–146. Springer, Cham (2023). https://doi.org/10.1007/978-3-031-42532-5_11
3. Ni, X., Samet, A., Cavallucci, D.: Build links between problems and solutions in the patent. In: Cavallucci, D., Brad, S., Livotov, P. (eds.) Systematic Complex Problem Solving in the Age of Digitalization and Open Innovation. TFC 2020. IFIP AICT, vol. 597, pp. 64–76. Springer, Cham (2020). https://doi.org/10.1007/978-3-030-61295-5_6
4. Berdyugina, D., Cavallucci, D.: Exploitation of causal relation for automatic extraction of contradiction from a domain-restricted patent corpus. In: Nowak, R., Chrząszcz, J., Brad, S. (eds.) Systematic Innovation Partnerships with Artificial Intelligence and Information Technology. TFC 2022. IFIP AICT, vol. 655, pp. 86–95. Springer, Cham (2022). https://doi.org/10.1007/978-3-031-17288-5_8
5. Trapp, S., Warschat, J.: LLM-based Extraction of Contradictions from Patents (2024). arXiv preprint arXiv:2403.14258
6. Livotov, P.: Nature's lessons, AI's power: sustainable process design with generative AI. Proc. Des. Soc. **4**, 2129–2138 (2024)
7. Jiang, S., Luo, J.: AutoTRIZ: Artificial Ideation with TRIZ and Large Language Models (2024). arXiv preprint arXiv:2403.13002
8. Wang, B., et al.: A task-decomposed AI-aided approach for generative conceptual Design. In: International Design Engineering Technical Conferences and Computers and Information in Engineering Conference (Vol. 87349, p. V006T06A009). American Society of Mechanical Engineers, August 2023
9. Kahneman, D.: Thinking, Fast and Slow. Macmillan, New York (2011)
10. Nori, H., King, N., McKinney, S.M., Carignan, D., Horvitz, E.: Capabilities of gpt-4 on medical challenge problems (2023). arXiv preprint arXiv:2303.13375
11. Guarino, G., Samet, A., Cavallucci, D.: PaTRIZ: a framework for mining TRIZ contradictions in patents. Expert Syst. Appl. **207**, 117942 (2022)

Challenges in Inventive Design Problem Solving with Generative AI: Interactive Problem Definition, Multi-directional Prompting, and Concept Development

Pavel Livotov$^{(\boxtimes)}$ ⬇ and Mas'udah ⬇

Offenburg University of Applied Sciences, Badstr. 24, 77652 Offenburg, Germany
`pavel.livotov@hs-offenburg.de`

Abstract. This paper explores the application of generative AI for systematic and inventive problem-solving in engineering design. Utilizing a multi-directional prompting approach, the study investigates the ability of AI chatbots to generate and evaluate innovative concepts based on numerous elementary solution principles. The research involved two sets of experiments with graduate and undergraduate students solving seven engineering design problems. The findings indicate that while generative AI can quickly generate a large number of ideas, it often overestimates the feasibility and usefulness of its solutions and tends to create overly complex designs. Comparisons of AI evaluations with those conducted by human participants revealed significant differences, highlighting the need for human oversight to ensure practical and contextually relevant outcomes. The experiments also revealed performance differences among various AI models, confirming a bias in self-assessment. Despite these challenges, integrating generative AI with multi-directional prompting using elementary inventive principles, primarily based on TRIZ methodology, proved effective in fostering innovative solutions. The study highlights the potential of simultaneously applying different AI models alongside human expertise to leverage the strengths of both large language models and human technical creativity.

Keywords: Generative AI · Autonomous Innovation · Inventive Design · Prompt Engineering · Engineering Creativity

1 Introduction

In recent years, generative artificial intelligence (AI) has attracted significant attention in various engineering domains due to its ability to autonomously generate content, facilitate ideation, solve complex problems, and assist in decision-making processes. In engineering design, generative AI holds great promise for enabling rapid prototyping, optimizing designs and streamlining iterative processes. However, to fully realize the potential of generative AI in engineering innovation, it is imperative to explore new methods that can creatively and autonomously solve engineering problems that go

© IFIP International Federation for Information Processing 2025
Published by Springer Nature Switzerland AG 2025
D. Cavallucci et al. (Eds.): TFC 2024, IFIP AICT 735, pp. 205–226, 2025.
https://doi.org/10.1007/978-3-031-75919-2_13

beyond traditional paradigms and professional expectations. The application of generative AI in the innovation process has been widely reported by numerous researchers. Recent advancements in generative AI have significantly impacted the field of engineering design, offering new tools and methods for solving complex problems. Deep Learning models, such as Generative Adversarial Networks (GANs), Variational Autoencoders (VAEs), and Large Language Models (LLMs), have shown promise in generating innovative outputs based on training data. A study by Brad [1] explores novel activation functions inspired by inventive principles to enhance the creative potential of these models, suggesting that human-behavior-inspired activation functions might increase their inventive capacity.

The integration of AI with TRIZ principles to handle multi-domain and heterogeneous data sources has also been investigated. Ayaou and Cavallucci [2] propose a unified representation framework to formalize various knowledge sources, establishing stronger links between unrelated sources to uncover hidden solutions. This approach illustrates the potential of AI in systematizing knowledge for innovative problem-solving.

Human-AI collaboration remains a crucial aspect of complex problem-solving. Memmert and Bittner [3] highlight the opportunities and challenges of hybrid teams, emphasizing the importance of designing suitable research contexts for studying such collaboration. Similarly, Qiu and Jin [4] compare ChatGPT and fine-tuned BERT models, underscoring the nuanced integration of AI with human expertise to enhance design support systems.

Based on a literature review, Müller, Roth and Kreimeyer [5] outline that the main barriers to collaboration between product development and AI application development are primarily due to communication issues, such as a lack of standardized process models for the use of AI, a lack of documented use cases and best practices, and others. Zhu, Zhang and Luo [6] explore the use of GPT for early-stage design concept generation, showing good performance in generating novel and useful design concepts. Large Language Models have demonstrated their utility in complex system design. Gomez et al. [7] investigate their application in detailed design phases, revealing areas for future research to improve LLMs' effectiveness in engineering tasks. Similarly, Ege et al. [8] benchmark AI design skills through a prototyping hackathon, highlighting generative AI's potential and limitations in ideation and decision-making. The ChatGPT of the OpenAI company (https://openai.com/chatgpt/) can generate ideas and provide instructions, but executing these ideas necessitates human intervention. This raises questions about the optimal level of human involvement in decision-making and problem-solving. Excessive human engagement may bias results, while insufficient involvement could result in impractical or unfeasible designs. Additionally, the subjective interpretation of ChatGPT's often vague instructions can lead to variations in their understanding and execution [8]. The integration of behavioral science principles into AI-driven systems is explored by Van Rooy and Vaes [9], emphasizing the need for a holistic approach that considers cognitive and behavioral aspects of human operators. Boussioux, Lane, et al. [10] explore creative problem-solving through human-AI partnerships, specifically targeting sustainable circular economy business ideas. Their study demonstrates a scalable and cost-effective approach to enhancing the early stages of the innovation process by integrating human-AI collaborative solution-seeking.

Xu et al. [11] compare ChatGPT's performance with human evaluators in engineering design tasks, addressing disparities in judgment confidence and decision reasoning. Their study underscores the need for consensus on confidence levels between AI and human evaluators to improve decision-making processes. Theoretical and practical benefits of integrating LLMs in engineering design are discussed by Chiarello et al. [12]. They highlight the potential for LLMs to automate various aspects of design tasks, increasing efficiency and exploring more options, while maintaining a balance between computational efficiency and human-centric design values. Finally, the application of image generative AI in creating design inspiration boards is evaluated by Ranscombe et al. [13]. Their study compares the quantity, variety, accuracy, and ambiguity of AI-generated boards with traditional methods, providing some insights into the optimal application stages of image generative AI in the design process.

The studies [14] and [15] highlight AI's success in enhancing the quantity and variety of ideas during initial brainstorming sessions, solving technical problems in process engineering [14], and conducting interdisciplinary, function-oriented searches for solutions. In these contexts, AI integration typically supports the idea generation where detailed engineering design of final solutions is not required. However, a significant challenge arises when attempting to create novel solution concepts that combine several complementary ideas and delivering comprehensive solutions to multiple partial problems simultaneously. This complexity is particularly evident in fields such as mechanical engineering, where a detailed AI-generated solution concept often falls short. Designers need more than a textual description and require clear, precise technical instructions or drawings to implement these concepts effectively. Moreover, the current text-to-image functionality of generative AI tools tends to produce random outputs, which are generally unsatisfactory for engineering purposes. These limitations underscore the necessity for more advanced AI capabilities and AI-aided inventive design approaches that can bridge the gap between conceptual design and practical implementation, facilitating a seamless transition from ideation to detailed engineering design.

This paper advocates for an integrative approach to automated multidirectional prompt generation, utilizing solution principles derived from various methodologies such as design theory [16], Theory of Inventive Problem Solving TRIZ [17], biomimetics and nature inspired design [14], process intensification, green engineering, and other systematic innovation approaches [18]. The goal is to enhance the effectiveness of generative AI chatbots in facilitating ideation sessions, brainstorming, and problem-solving activities within the field of engineering design. Additionally, the paper analyzes strategies for AI-aided development of comprehensive solution concepts based on the integration of multiple ideas. It also critically examines the capability of AI to objectively evaluate and select the most robust ideas and concepts, ensuring their practical applicability in engineering design.

Through controlled experiments and analysis of different prompting strategies, this paper aims to identify patterns and correlations that indicate the effectiveness of different prompting approaches. It assesses how integrative, multi-directional prompting strategies enhance the creative potential of generative AI chatbots, particularly in scenarios where the chatbots operate autonomously or in a collaborative ideation group. However, the current stage of research may limit the generalizability of the findings beyond the proposed prompting strategies, existing AI chatbot architectures, underlying large language models, and the specific technical problems and tasks addressed. Additionally, the objective assessment of engineering creativity presents challenges and may not fully capture the diverse and nuanced criteria of creative output.

The findings of this research can be valuable for both academia and industry, providing insights into optimizing generative AI systems for technical innovation. By highlighting the nuances of prompting strategies and their impact on creativity, this study aims to support the development of more effective AI chatbot-based tools for engineering design, problem-solving, and ideation processes.

2   Multidirectional Prompting

There are a number of different approaches to the formulation of prompts for generative AI chatbots. AI-based idea generation typically uses zero-shot or few-shot settings for prompt generation. In a zero-shot setting, the AI model is given prompts and asked to generate responses without any prior examples or specific training on the task at hand. This relies on the model's broad generalization capabilities, derived from extensive pre-training on diverse datasets. In contrast, in a few-shot setting, the AI model is provided with a small number of examples related to the task to guide its responses. This method uses the examples provided to improve the model's understanding and generate contextually relevant and more accurate results.

This paper proposes the concept of Multidirectional Prompting (MDP), which is defined as the application of various elementary solution principles or stimuli to generate ideas and innovative solution concepts using generative AI chatbots. Innovative or inventive solution concepts are characterized by the novel, useful and feasible combination of one or more solution ideas tailored to the specific problem or goals at hand. In the MDP approach, the search for solution ideas takes place simultaneously in several directions. These directions are determined by the defined sub-problems and sets of different inventive stimuli selected by the user. This comprehensive approach allows the AI to explore a wide range of potential solutions, addressing different facets of the problem simultaneously. By integrating different elementary solution principles, the MDP method aims to enhance the generative capabilities of the AI, promoting the development of more holistic and effective solutions.

The multidirectional prompting (MDP) can be carried out using various strategies tailored to the given problem. These strategies include:

Systematic Approach. In this method, the chatbot is given a predefined number of solution principles (SPs), for example ten. For each SP, the chatbot generates a specific number of ideas, for instance, five ideas per SP. The SPs are applied independently, one after the other. After the initial pool of ideas is created (e.g., 10 SPs × 5 ideas each = 50 ideas), the chatbot then uses just one idea or combines several complementary ideas (two or more) into innovative solution concepts, for instance, proposing five or more comprehensive solution concepts.

Random Approach. In this method, the chatbot is provided with a certain number of solution principles and applies them all simultaneously, without a predefined order. The chatbot generates solution concepts by applying one or more solution principles in a single step, resulting in a number of solution concepts.

Collaborative Systematic Strategy. This strategy involves two or more chatbots, potentially utilizing different large language models (LLMs) like ChatGPT and Google Gemini, working together in a session. Initially, each chatbot generates ideas using the systematic approach. In the second step, these chatbots exchange their generated ideas to create solution concepts. Essentially, each chatbot leverages ideas from the other chatbot(s) to develop comprehensive solution concepts.

Multiple Problem Situations. In complex scenarios, the problem can be divided into a set of independent or interdependent sub-problems with varying priorities or rankings. The problem-solving process begins with addressing a core problem, typically the highest-ranking sub-problem, followed by solving other lower-ranking sub-problems either sequentially or in parallel. In this context, chatbots must identify and rank the sub-problems, generate ideas for each sub-problem, and ultimately create a solution concept that ideally integrates complementary solution ideas for all sub-problems.

This paper employs a method for the automated formulation of elementary creative stimuli for product or process design at various levels of abstraction and across different engineering domains [15]. This method, proposed and validated by the authors in both industrial and educational settings, has demonstrated its efficacy in generating innovative solutions and enhancing the design process. The knowledge base for systematic ideation is primarily built upon 160 elementary inventive principles, enhanced and extended by selected TRIZ tools. These include the 40 inventive principles, trends of technical evolution, standard solutions, and selected physical, chemical, biological, and geometrical effects. Additionally, the knowledge base incorporates methodologies from other disciplines such as biomimetics, process intensification, and green engineering [18]. To maintain efficiency, overlapping or redundant classical inventive operators known in TRIZ were excluded. Automated idea generation can be conducted at various levels of problem analysis and understanding, such as:

- Improving or transforming the system or its components,
- Providing or increasing the useful action,
- Eliminating harmful effects or undesirable characteristics,
- Resolving engineering contradictions, where improving one target parameter causes another target parameter to deteriorate.

An additionally developed application for automated prompt generation has access to 200 predefined elementary inventive principles and various engineering domains for solution search. Users can further extend this database with their own specific preferences. The application operates through four key steps:

(1) **Pre-Selection of Elementary Inventive Stimuli and Engineering Domains**. In this step, users pre-select elementary inventive principles and engineering domains based on the initial problem situation. The number of applied principles is not restricted, and additional principles can be introduced at any phase of the process. For instance, in mechanical engineering design problems, users often start with design and purely

mechanical inventive principles and later extend their selection to include princi-ples from other engineering domains, such as acoustic, thermal, electromagnetic, and others, as illustrated in Fig. 1. Conversely, in process engineering challenges, universal inventive principles that are independent of any specific technical domain are typically applied first. Users also have the flexibility to review all principles and create a tailored set of inventive stimuli based on their own experience, intuition, or other criteria.

(2) **Interactive Problem Definition**. In this initial phase, both the AI chatbot and the users collaboratively discuss the problem situation until a comprehensive and precise problem definition is achieved. It is crucial to consider all relevant information and constraints while avoiding unwanted biases from less relevant secondary issues. As a result, the AI chatbot delivers an improved version of the problem definition and highlights key data points that demonstrate its correct understanding of the initial problem situation. This includes aspects such as the ideal final result, the active and passive components in the system, positive functions, and undesirable effects or harmful properties to be eliminated. If needed, additional problem analyses, such as cause-effect chain analysis for identifying root causes of the problem, can be conducted. This further refines the problem definition and ensures a thorough clarification and understanding of the underlying issues.

(3) **Step-by-Step Idea Generation by AI Chatbot.** Once the user confirms the start of the ideation process, the chatbot generates a specified number of ideas, such as, for instance, five ideas per solution principle, applying each solution principle independently, one after the other. This process is user-controlled, allowing the user to interactively request more ideas, shift to the next solution principle, or interrupt the ideation process to proceed to the final phase of concept creation. This iterative and flexible approach ensures that the ideation process remains aligned with the user's goals and preferences, fostering the generation of diverse and innovative ideas.

(4) **Solution Concept Creation.** In this critical and often less explored step, the AI chat-bot combines different ideas into various innovative solution concepts, typically gen-erating five, ten, or more concepts. The goal is to optimally meet the defined inventive objectives by leveraging the broad generalization capabilities of the underlying large language model, which has been extensively pre-trained on diverse datasets. This step synthesizes the previously generated ideas into comprehensive and consistent solutions, ensuring that the final concepts align with the specified inventive goals and address the problem effectively. Depending on the outcomes, the user can inter-actively instruct the AI to follow different concept creation strategies. These strate-gies might focus on criteria such as highest value and usefulness, high novelty and patentability, high feasibility with low implementation efforts, the use of a specific technology or engineering domain, or selected core ideas. This interactive approach allows the user to tailor the solution concepts to better fit specific requirements or constraints, enhancing the overall effectiveness and applicability of the generated solutions.

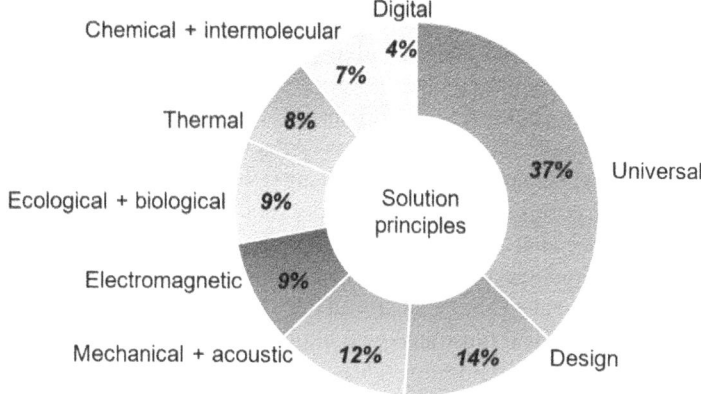

Fig. 1. Variety of the elementary solution principles for multidirectional prompting.

3   Experimental Procedure

This section describes a series of experiments investigating the application of AI-based automated idea and solution concept generation to engineering design problems. These experiments were conducted using student projects from a one-semester course on AI-based inventive design and new product development at the University of Applied Sciences Offenburg, Germany. Five groups of graduate students and a group of the undergraduate students from different years of study participated in the experiments. Additionally, the experiments were conducted by a control group comprising the authors of this study. The undergraduates were in their 5th or 6th semester and were studying subjects unrelated to mechanical engineering. The graduate students were in their 8th or 9th semester and studying for a Master's degree in Mechanical Engineering and Robotics. All students received basic instruction from the same tutor in systematic new product development and in solving inventive problems using the inventive principles of the TRIZ methodology. Students worked in groups of 2–3 and used generative AI chatbots, enhanced by automated multidirectional prompt generation based on elementary inventive principles. This approach aimed to facilitate systematic and innovative problem solving in their respective design engineering challenges.

The experimental part of the research study consisted of two series of experiments. In the first series, all groups received the same pre-formulated problem and the same initial prompt for use with generative AI chatbots. The comprehensive multidirectional prompt consisted of the following major sections and task assignments:

(1) Problem Description and Refinement. The AI chatbot was tasked with confirming and refining the given problem formulation. During this phase, the user had to interactively confirm or correct the chatbot's proposals until the problem formulation was completed. Only after this confirmation could the idea generation stage be initiated.

(2) Selection of Elementary Inventive Principles. This step involved the multidirectional prompted systematic search for solutions. The initial number of ideas for each solution principle was also defined. For the first experiment, 10 design solution principles were selected, with at least 5 ideas generated for each principle. Consequently, at

least 50 solution ideas should be proposed by the AI Chatbot in one session. Unique ideas must be presented in a table, avoiding repetitions, and should include a unique idea ID number and name, a description of the idea, a separate implementation example, and the corresponding solution principle.

(3) Creation of Innovative Solution Concepts. After utilizing all the solution principles in the idea generation process, the chatbot was asked to create at least five innovative solution concepts. These concepts should combine and enhance different complementary ideas to effectively address the problem. Each solution concept should offer a comprehensive approach, incorporating various ideas to enhance useful actions and prevent or mitigate harmful effects. The solution concepts should be thoroughly described and illustrated with implementation examples, demonstrating the integration of different inventive ideas into practical, workable solutions.

(4) Solution Concept Evaluation. This final prompt section involves assigning the chatbot to evaluate its solution concepts based on the following criteria, using the initial scales applied in the experiments:

- Feasibility (0 - unviable for implementation; 1 - feasible but requires substantial effort; 2 - easily implementable),
- novelty (0 - common or derivative; 1 - moderately novel; 2 - highly original or unique),
- usefulness (0 - irrelevant for the problem; 1 - moderately relevant; 2 - highly useful).

In the first series of experiments the engineering design problem "Twist-off Screw Cap" was defined in the prompt as follows: "Many foods are sold in jars or bottles with screw caps (twist-off caps). The thread on the mouth of the jar or bottle means that a screw cap can be closed tightly during filling and re-closed several times after opening. A plastic sealant is injected into the inside of the cap to ensure a secure seal. During the filling process, the screw caps are screwed onto the jars with high torque to ensure that the food is stored safely for a long time. Opening the jar is made difficult by the resulting friction in the pre-tensioned thread of the screw cap. In addition, once the jar is closed, the food is under vacuum. The vacuum increases the friction between the cap and the jar, making it even more difficult to open the screw cap. As a result, consumers often find it difficult to open jars. We are looking for solutions that make it easier to open jars closed with twist-off screw caps without additional tools and without disadvantages for manufacturers and consumers."

As an example, in the interactive problem formulation mode, ChatGPT3.5 has extended the problem formulation as follows: "The goal is to develop solutions that make it easier to open these jars without requiring additional tools and without introducing disadvantages for manufacturers or consumers. The selected problem definition categories are: a) Work Tool/Working Medium: the screw cap equipped with a thread and an internal plastic sealant; b) Target Objects: The jars or bottles being sealed; c) Main Useful Action: Securely sealing the jars or bottles to preserve the food, ensuring it is safe and has a long shelf life; d) Harmful Effect to be Eliminated: The excessive difficulty for consumers in opening the jars due to high friction caused by the torque and vacuum seal."

Figure 2 on the left illustrates this initial problem situation, while on the right shows the control solution comprising a two-piece twist-off lid for glass storage jars, according to the US Patent US6662958B2 (2003). This control solution features a cap made of two separate parts: the first part is a sheet metal covering disc (12) with an elastic sealant that covers the edge of the jar neck (11), and the second part is a cylindrical threaded ring (32). The inner surface of the threaded ring has retaining thread members (40) for locking the disc (12) onto the jar neck (11) and for unlocking the cap by lifting the disc (12) with the bead (34) while unscrewing the ring. This means the user only needs to overcome the thread friction initially when opening the jar.

The control solution was not disclosed to the participants of the experiments. During the evaluation phase, we assessed how closely the solutions generated by the systematically guided and prompted generative AI chatbots, as well as those created by the student teams, matched the control solution. This evaluation aimed to measure the effectiveness of the AI-driven ideation process and the students' problem-solving abilities in approximating the control solution.

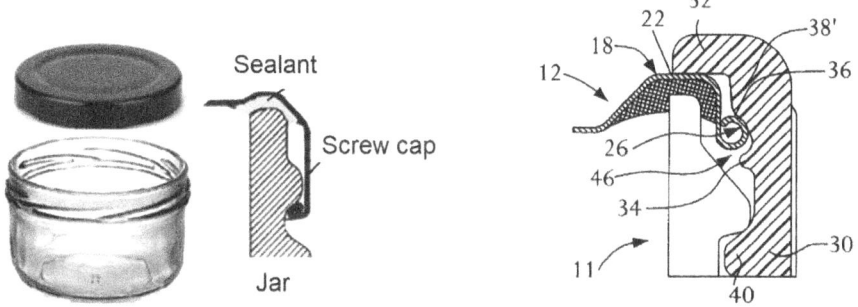

Fig. 2. The existing twist-off cap to be improved (on the left) and the control solution (on the right) as a two-piece twist-off cap as described in US Patent US6662958B2 (2003).

The 10 design elementary solution principles out of a total of 200 principles for idea generation by AI chatbots, predefined in the prompt, are presented in Table 1. It should be emphasized that the ID numbers of the elementary solution principles used in the prompts do not correspond to the ID numbers of the 40 TRIZ Inventive Principles. However, many of the 200 elementary solution principles correlate with the sub-principles of the 40 TRIZ Inventive Principles.

Table 2 illustrates exemplarily 5 ideas proposed by the ChatGPT 4.0 for the Solution Principle 7 "Segment the screw cap with a sealant into several independent modules, parts or sections; design the screw cap to be dismountable". Notably, some ideas, such as the 11. Modular Cap Design, closely resemble the control solution. However, the descriptions lack sufficient detail for users to fully recognize the control solution. This phenomenon illustrates a significant challenge in applying TRIZ inventive principles or generative AI prompted with these principles for solving engineering design problems.

Table 1. Ten solution principles for engineering design problems applied in multi-directional prompt for the work tool "Screw cap with a thread and internal sealant".

ID-Nr.	Abstract description of the solution principle
#2	Change mechanical or surface properties of Work Tool like strength, hardness, density, viscosity, roughness, wettability etc.
#6	Pre-arrange Work Tool so it can come into action at the most convenient position and without losing time
#7	Segment the Work Tool into several independent modules, parts or sections; design the Work Tool to be dismountable
#8	Replace the Work Tool by several smaller units, with same or different working parameters
#16	Remove the disturbing parts or substances from the Work Tool responsible for harmful effect
#21	Divide Work Tool into elements which position can be changed relative to one another
#22	Use Work Tool with adaptive and flexible elements, like joints, springs, elastomers, fluids, gases
#30	Design Work Tool with multi-layered structure
#34	Use oscillating or vibrating Work Tool
#41	Apply flexible shells, thin films or membranes in Work Tool design

The gap between abstractly formulated design solutions in text and their graphical implementation by designers is substantial, making it difficult to translate textual descriptions into practical, visualized solutions.

The authors advocate for a multi-directional prompting approach that generates a substantial number of abstract solution ideas based on numerous selected solution principles. The experiments aim to explore the benefits and potential of AI in rapidly analyzing these ideas and rapidly generating robust solution concepts, including complementary and synergetic ideas. This approach seeks to address the current limitations of AI in inventive engineering design, which often relies on direct suggestion methods that attempt to create innovative solutions from a single or few ideas. The experiments also aim to assess the reliability and objectivity of AI in autonomously selecting the strongest ideas from a large pool, evaluating the resulting solution concepts, and recommending the best concepts for further implementation and assessment. By systematically investigating these aspects in the context of multi-directional prompting, the study aims to improve to enhance the effectiveness of AI in the inventive engineering design process, bridging the gap between abstract textual formulations and practical graphical implementations.

The results of the first experimental series are presented and discussed in the Sect. 4. An additional purpose of the first series of experiments was to introduce participants to the formalized approach for multidirectional prompting. This phase aimed to familiarize students with the systematic method of generating ideas through predefined elementary solution principles. During this series, students received hands-on training on

Table 2. Examples of ideas proposed by the ChatGPT4.0 for the Solution Principle 7 "Segment the screw cap into several independent modules, parts or sections" in the control group.

Nr.	Idea Name	Idea Description	Implementation Example
11	Modular Cap Design	Create a cap with separate, removable sections that reduce opening torque collectively	Caps with top and side parts that detach or loosen separately
12	Split-Ring Cap	Develop a cap with a split-ring design allowing part of it to expand for easier opening	A cap with a segment that unfolds or expands when twisted
13	Hinged Cap	Introduce a cap with a hinged section that can flip open to break the seal easily	Caps with a built-in lever that when flipped, opens the cap
14	Independent Thread Rings	Design the cap with independent thread rings that can move slightly to reduce friction	Multi-ring caps where rings can rotate independently
15	Detachable Seal System	Implement a cap where the sealant layer is a separate, removable module	Caps with an easily removable and replaceable seal layer

how to interact with generative AI chatbots, refine problem definitions, select and apply elementary inventive principles to generate innovative solutions ideas and concepts.

The training included step-by-step guidance on confirming and refining problem formulations, selecting appropriate inventive principles, and combining generated ideas into comprehensive solution concepts. By the end of the first series, students were expected to be skilled in using the AI tools and methodologies provided, equipping them for more autonomous and independent work on their own chosen engineering design problems in the second series of experiments. This preparation was crucial for ensuring that students could effectively apply the multidirectional prompting approach to diverse engineering problems of their choice without direct supervision.

Thus, in the second series of experiments, participants interactively defined their problems using AI, selected elementary inventive principles for idea generation, and proposed and tested different strategies for creating innovative concepts based on strong and complementary ideas. They also assessed the accuracy of AI-driven concept evaluations. Additionally, they compared the performance of different AI chatbots, such as OpenAI ChatGPT (https://openai.com), Google Gemini (https://gemini.google.com/), Anthropic Claude (www.anthropic.com) and others, both against each other and against their own personal creativity.

4   Experimental Outcomes and Discussion

This section briefly presents and discusses the primary outcomes from the conducted experiments. Due to the extensive nature of the study, not all results and comparisons can be included here. The authors acknowledge this limitation and plan to publish a comprehensive analysis of the full study in a subsequent publication. The detailed report will provide a thorough examination of the experimental data, comparative evaluations, and further insights into the application of multi-directional prompting and generative AI in inventive engineering design.

Illustrating concepts and ideas through AI-generated technical drawings was not an explicit part of this study. However, systematic attempts by experiment participants to generate technical drawings using AI have proven to be less effective. This process requires special iterative and time-consuming prompt formulations, and for most tasks, the resulting drawings are predominantly random, often overly detailed, and technically inaccurate. Consequently, these drawings were generally of limited practical value and could at best serve as a source of inspiration rather than precise engineering documentation.

4.1   Autonomous Solution Concept Creation by Generative AI

In the first series of experiments addressing the pre-defined "Twist-off cap" problem, generative AI chatbots were prompted with the same 10 inventive design principles outlined in Sect. 3. This process resulted in the automatic generation of 50 different ideas. In the next step, the AI chatbots were tasked with autonomously generating solution concepts using the following universal prompt, without any specific instructions or constraints:

> "… create 10 innovative solution concepts by combining different complementary ideas to address the problem. These solution concepts should offer comprehensive solution approaches to the problem, incorporating various ideas to enhance the useful action and prevent or mitigate harmful effects. Present the solution concepts in a table format: number including concept name; concept description; ID numbers of ideas used in the concept and implementation example. The implementation examples should illustrate how each solution concept can be implemented in practice, showing the integration of different inventive ideas into a workable solution".

Table 3 illustrates exemplarily 10 solution concepts proposed by the ChatGPT 4.0 in the control group comprising the authors of the study. The concept name, concept description, selection of ideas used in the concept, order of cited ideas, and the elementary solution principles for idea generation were suggested exclusively by the generative AI chatbot.

As observed, only 20 out of the 50 ideas proposed during the ideation phase, were used in the solution concepts of the control group. (This phenomenon was also observed in all teams in the first series of experiments, with an average of 21.4 out of 50 ideas (43%) generated with ChatGPT 3.5 and 4.0.) Notably, five ideas (11. Modular Cap Design,

37. Thermal Insulation Cap, 41. Vibratory Cap, 46. Flex-Shell Cap, 49. Breathable Membrane Cap) are used twice across different concepts. This repetition suggests that these ideas may possess a higher inventive level or greater potential for solving the given problem. However, when ChatGPT was asked to select the top 5 ideas out of 50 based on usefulness, the resulting ideas were: 7, 22, 31, 34, and 50. Except for idea 7, these ideas are mentioned only once in the concepts (see Table 3). The idea 7 (One-Touch Release Cap), which proposes partial unscrewing with a single press to reduce initial friction in the seal, is not included in any concept. However, idea 7 is quite similar to idea 31 (Spring-Embedded Cap), which incorporates springs within the cap to automatically reduce the initial opening force.

As noted by the authors in numerous experiments, generative AI chatbots often deviate from the proposed elementary solution principles during the idea generation phase. This occurs in both multi-directional prompting and when using a single solution principle in a prompt. For example, idea 9 (Magnetic Positioning Cap), and idea 37 (Thermal Insulation Cap), which suggests using heat to loosen the seal, were generated despite the absence of magnetic or thermal solution principles in the prompt.

Similar to the analysis of the frequency of use of the same ideas in the solution concepts, it can be analyzed if the underlying elementary solution principles, presented in Table 1, were used more frequently, assuming that these solution principles have a higher probability of success in the search for feasible and useful solutions. As shown in Table 3, the 10 generated concepts use principles #41 (Flexible shells, thin films) - 5 times, #22 (Adaptive and flexible elements) - 4 times and #34 (Oscillating or vibrating work tool) - 3 times. Principles #7 (Segment Work Tool into independent parts), #8 (Replace Work Tool by several smaller units) and #30 (Multi-layered structure) are used 2 times each and principles #6 (Pre-arrange Work Tool), #16 (Remove the disturbing parts), #21 (Work tool division into elements with variable position relative to each other) only once. It is important to note that the underlying solution principles #7, #21 used in the control solution of the two-piece cap are not at the top of this list. Furthermore, the ideas based on recommended solution principle #2 (Changing mechanical or surface properties) were not even included in the control group concepts.

In order to verify this situation more objectively, ChatGPT4.0 was asked in a separate chat to identify the 10 most promising principles out of 88 solution principles defining different inventive approaches to transform or improve the work tool (Screw cap with sealing) to effectively solve the Twist-Off Screw Cap problem described above. According to ChatGPT, the following 10 solution principles appear to be the most promising for developing workable solutions to the jar-opening process: #2, #6, #7, #20 (Adjustable Work Tool), #22, #23 (Periodic or pulsed action), #41, #45 (Use gas or liquid as working elements), #53 (Change the shape and geometry), #58 (Composition of materials in different aggregate states). It is noteworthy that the choice of solution principles made by the generative AI corresponds to 50% of the inventive design principles chosen by the authors for the Twist-Off Screw Cap problem. Furthermore, all of the top 10 solution principles selected by ChatGPT belong to the domain of design or mechanical engineering.

Additionally, it cannot be excluded that during the ongoing concept development phase, there may be a correlation between formulated concepts, resulting in a biasing effect. This bias could lead to a lower variety of autonomously proposed concepts, thus limiting the diversity of the final solutions.

Table 3. Examples of 10 solution concepts with corresponding ideas and underlying elementary solution principles (Table 1) proposed by the ChatGPT 4.0 in the control group.

No.	Concept name	Concept description	Ideas applied in concept	Solution principles
1	Adaptive Ease Cap	Combines a flexible shell, a self-healing film that softens during opening, and a vibratory mechanism for ease of opening	46. Flex-Shell Cap 50. Self-Healing Film Cap 41. Vibratory Cap	#41 #34
2	Smart Seal Cap	Utilizes a breathable membrane to equalize pressure and a fluid-filled seal for dynamic friction reduction, enhancing usability	49. Breathable Membrane Cap 33. Fluid-Filled Cap Seal	#41 #22
3	Multi-Layer Twist Cap	Features multiple layers including thermal insulation and acoustic feedback for user-friendly, intuitive opening	37. Thermal Insulation Cap 38. Acoustic Feedback Layers	#30
4	Modular Release System	A cap design with independent modular sections and an embedded elastic film, allowing each part to move or stretch individually	11. Modular Cap Design 48. Elastic Film Cap Liner	#7 #41
5	Ultimate Release Cap	Integrates spring-embedded sections, a magnetic positioning system, and an oscillating grip to provide the easiest opening	31. Spring-Embedded Cap 9. Magnetic Positioning Cap 42. Oscillating Grip	#22 #6 #34

(*continued*)

Table 3. (*continued*)

No.	Concept name	Concept description	Ideas applied in concept	Solution principles
6	Comfort Grip Cap	Combines the flexible shell with the modular cap design and multi-part system for enhanced grip and ease of opening	46. Flex-Shell Cap 11. Modular Cap Design 16. Multi-Part Cap System	#41 #7 #8
7	Pressure Equalizer Cap	Utilizes gas-pressure equalization and a breathable membrane to automatically adjust internal pressure, making the cap easier to open	34. Gas-Pressure Equalizer Cap 49. Breathable Membrane	#22 #41
8	Thermal Twist Cap	Features a thermal insulation cap with embedded vibratory mechanisms that activate through touch, simplifying the initial twist action	37. Thermal Insulation Cap 41 Vibratory Cap	#30 #34
9	Segment Flex Cap	Integrates segmented grip enhancers with flexible cap bands and adaptive thread tracks, distributing the force required to open the cap	19. Segmented Grip Enhancers 32. Elastic Bands 27. Adjustable Thread Tracks	#8 #22 #21
10	Ultimate Safety Seal Cap	Combines frictionless inner lining, non-adhesive sealant, and a vacuum release button for a smooth and effortless jar opening experience	21. Frictionless Inner Lining 23. Non-adhesive Sealant 22. Vacuum Release Button	#16

Finally, the comparison of the concepts presented in the given example shows that none of them could come close to the two-piece twist-off cap disclosed in US Patent US6662958B2 (Fig. 2). Only the solution concepts 4 (Modular Release System) and 6 (Comfort Grip Cap), which incorporate ideas 11 (Modular Cap Design) and 16 (Multi-Part Cap System), respectively, can be considered as having the potential to approximate the control solution. An analysis of all experiments performed by students in the first experimental series reveals that no concepts proposed autonomously by generative AI possess features similar to the patented two-piece cap. This observation highlights a significant weakness of generative AI in solving inventive design problems at their current stage of development. All experiments were conducted between April and June 2024. In general, the use of a control solution, defined as a feasible design solution to be achieved through AI-assisted problem solving, proved to be an effective method for objectively verifying the quality and progress of generative AI in inventive engineering design.

4.2   Limitations of AI-Powered Evaluation of Innovation Concepts

The methodology for the AI-assisted evaluation of innovation concepts was described in detail in Sect. 3. To ensure objective and comparable results, the evaluation of the concepts for feasibility, novelty and usefulness was conducted in separate chat sessions. Each session included the problem definition and a detailed description of the concepts along with their implementation examples. This evaluation aimed to address the following research questions:

a) How does evaluation by generative AI chatbots compare to evaluation by experimental participants?
b) How does the evaluation by generative AI chatbots compare to the evaluation by other AI chatbots using different Large Language Models?

In both experimental series I and II, the generative AI chatbots tended to overvalue their own concepts compared to the ratings provided by the participants. Typically, the AI chatbots assigned significantly higher values for usefulness and feasibility than those given by the participants, as shown in Table 4 for the concepts created by ChatGPT4.0 in the control group during the first experimental series. The additional use of finer scales, e.g. 5- and 10-point rating scales, for idea evaluation in the second set of experiments did not substantially change this observed overestimation effect, which was particularly pronounced with ChatGPT.

It should be noted that the evaluation of concepts by the same AI chatbot can vary slightly, typically by plus or minus one point, when a repeated evaluation is requested within the same chat or in a separate chat session. Additionally, the rating of individual concepts may differ slightly from the bulk or aggregate rating of multiple concepts evaluated simultaneously. Experiments have shown that using finer scales, such as a 10-point scale, can help compensate for these variations in the AI chatbots' ratings, providing a more nuanced and consistent assessment.

The objective evaluation of solution concepts by generative AI chatbots is crucial in the context of an automated or autonomous AI-supported inventive design process. Thus, a focus of the experiments was also the comparison of concept evaluations performed by different AI chatbots using different Large Language Models. Interestingly, in a pairwise comparison, both ChatGPT4.0 and Gemini rate the usefulness of their own concepts higher. For example, in the control group, ChatGPT scores 9 out of its 10 concepts (i.e. 90%) as "2 - highly useful" and one as "1 - moderately relevant". Gemini rates only 70% of ChatGPT concepts as "2 - highly useful". Conversely, Gemini rates 100% of its concepts as "2 - highly useful", while ChatGPT rates 70% of Gemini concepts as "2 - highly useful" and the rest as "1 - moderately relevant".

Table 4. Example of the self-evaluation of solution concepts generated by the ChatGPT4.0 compared to the evaluation of the experiment participants (control group) in the first series of experiments.

No.	Concept name	Feasibility		Novelty		Usefulness	
		ChatGPT	User	ChatGPT	User	ChatGPT	User
1	Adaptive Ease Cap	1	1	2	1	2	2
2	Smart Seal Cap	1	0	2	1	2	1
3	Multi-Layer Twist Cap	1	1	2	1	2	1
4	Modular Release System	1	1	2	2	2	1
5	Ultimate Release Cap	1	0	2	2	2	1
6	Comfort Grip Cap	1	1	1	2	2	1
7	Pressure Equalizer Cap	1	0	2	2	2	1
8	Thermal Twist Cap	1	0	1	2	1	1
9	Segment Flex Cap	1	1	1	2	2	1
10	Ultimate Safety Seal Cap	1	1	2	2	2	2
	Total score	10	6	17	17	19	12

Gemini is also more critical of concept novelty than ChatGPT, rating only 30% of ChatGPT concepts as "2- highly original or unique". In contrast, ChatGPT rates 70% of its concepts as "1 - highly original". However, Gemini is more cautious about the novelty of its own concepts, with only 20% "2 - highly original" and 80% "1 - moderately novel". It is noticeable that both ChatGPT and Gemini never consider their own concepts as unfeasible, but the concepts of others. For example, ChatGPT rates only 10% of Gemini 's concepts as "2 - easy to implement", 80% as "1 - feasible with considerable efforts", and 10% as "0 - unfeasible". On the contrary, Gemini rates 50% of its own concepts as 50% easily implementable and 50% feasible with efforts.

It appears that generative AI models show a moderate Not Invented Here effect in the evaluation of engineering solutions. This more critical attitude to concept evaluation was observed across all experiments. The authors consider this phenomenon beneficial for achieving a more balanced evaluation of ideas and concepts by different AI chatbots acting as virtual teams of specialists autonomously or together with the engineers.

4.3 Summary of the Second Experimental Series

In the second set of experiments, six groups of students experienced in using generative AI for inventive problem solving were given the opportunity to apply their skills and engineering creativity to a design problem of their choice. Participants were responsible for defining the problem, selecting appropriate elementary solution principles for multi-directional prompting, and applying them within generative AI chatbots of their choice. They were also free to define their concept generation approaches, including predefining promising combinations of generative solution principles, identifying the strongest ideas, and setting other criteria. For example, focusing on developing solution concepts around a single core idea can provide more targeted improvements. Table 5 shows the scope of the second set of experiments for the six student groups, presenting the number of elementary solution principles selected by the students, the number of ideas generated by the AI chatbots, the number of solution concepts created and evaluated by the AI, and the applied generative AI chatbots.

Table 5. Experimental scope in the second series of experiments using different AI chatbots, like ChatGPT 3.5/4.0/4o, Google Gemini and Anthropic Claude.

Gr.	Design problem name	Number of			AI tools applied	Concept creation strategy
		solution principles	generated ideas	solution concepts		
1	Smoke detector	10	50	10	ChatGPT, Gemini	a) autonomous proposal by AI
2	Hot drink cup	10	45	12		b) based on 10 strongest AI ideas selected by AI
3	Quick release chuck	14	50	30	ChatGPT, Gemini, Claude	
4	Shape adaptive gripper	10	196	18		c) based 1…10 strongest AI ideas selected by the users
5	Barbeque grill	22	150	33		
6	Cable winder	10	150	30		

Table 6 presents the results of an anonymous survey conducted among the students at the end of the second series of experiments. The 17 experiment participants responded to the survey questions using the following 10-point scale: 1–2 (very low), 3–4 (low), 5–6 (medium), 7–8 (high), and 9–10 (very high).

Here is a summary of the observations and recommendations from the experience of the groups in the experiment.

(1) Combining AI-powered and human-driven problem definition, ideation, concept creation, and evaluation is a balanced and effective approach. This collaboration leverages the strengths of both AI and human expertise.

(2) Generative AI can exhibit hidden preferences, introducing unexpected biases that are often challenging to recognize. Therefore, monitoring and adjusting AI-driven innovation processes are essential to achieving workable and inventive solutions. In light of the current state of AI technologies, human judgment is critical for maintaining focus and addressing subtle innovation aspects.

Table 6. Results of an anonymous survey on the performance of generative AI in the second series of experiments with a scale 1–2 (very low), 3–4 (low), 5–6 (medium), 7–8 (high), and 9–10 (very high).

No.	Survey question: How do you rate the following aspects in application of generative AI …	Groups of participants (mean values)						
		1	2	3	4	5	6	all groups
1	contribution of AI to increasing your personal inventive CREATIVITY?	7.3	8.0	6.3	8.0	6.3	7.0	7.1 SD = 1.6
2	performance of AI in terms of the ideas USEFULNESS?	7.5	6.5	6.3	6.0	7.0	3.0	6.1 SD = 2.1
3	performance of AI in terms of the ideas NOVELTY?	5.8	7.0	8.7	7.5	6.3	5.7	6.7 SD = 1.9
4	performance of AI in terms of the ideas FEASIBILITY?	6.8	4.5	3.7	3.0	5.3	3.7	4.7 SD = 1.9
5	overall performance of AI in the Solution Concept Development phase?	7.8	7.5	5.7	5.5	6.7	3.0	6.1 SD = 2.0
6	level of detail of the solution concepts proposed by the AI, so that designers can quickly implement a solution concept?	5.5	6.0	4.0	3.0	6.7	2.7	4.7 SD = 2.0
7	accuracy of the evaluation of solution concepts by AI?	4.3	5.0	3.7	4.5	3.3	4.7	4.2 SD = 1.7

(3) Creating accurate and useful technical drawings for visualizing ideas and solution concepts remains one of the most significant challenges in the path to autonomous design problem-solving.

(4) AI tends to overrate its own ideas and concepts, particularly in terms of usefulness and feasibility, and often develops too complex solutions. AI-based evaluations can vary slightly with repeated requests and differ between individual and bulk evaluations. Currently, AI is not suitable for objectively evaluating technical concepts.

(5) The varying performance of different AI tools presents an opportunity to create mixed AI teams for problem-solving. Tools such as OpenAI ChatGPT, Google Gemini, and Anthropic Claude show substantial differences in performance across various steps of the problem-solving process, including idea generation, concept creation, and evaluation. These tools can complement, monitor, and, if necessary, correct each other. Most AI-generated ideas and solutions across different chatbots are often similar, though each chatbot also offers some unique solutions.

(6) The most stable performance in using automatically generated multi-directional prompt instructions is currently achieved with ChatGPT and Anthropic Claude. The free version of Google Gemini may deviate from given instructions and often requires

user interaction for precise prompt adherence. Nonetheless, Gemini tends to provide more objective evaluations of solutions.

(7) Multi-directional prompting with elementary solution principles enables AI to quickly generate a vast number of ideas and identify the strongest ones to create a smaller set of solution concepts. This approach opens up new perspectives despite the large number of non-useful ideas. Although AI-generated solution concepts are often over-engineered or infeasible, they can still inspire practical solutions.

5   Concluding Remarks and Outlook

The results of this study highlight several key findings and implications for the application of generative AI in inventive engineering design.

First, using the multi-directional prompting of AI chatbots with elementary solution principles can dramatically increase the productivity and variety of idea generation, significantly outperforming engineers using classical creative brainstorming-based methods or systematic idea generation approaches such as TRIZ. The main challenge therefore shifts to selecting the strongest ideas and creating and evaluating solution concepts based on these strong ideas. The research task is to develop and evaluate various concept development strategies for both fully autonomous and interactive searches for inventive design solutions facilitated by AI. An important question remains to determine the optimal level of human involvement in AI-assisted inventive problem solving.

Second, the tendency of AI chatbots to overestimate the usefulness and feasibility of the concepts they generate suggests a need for improved self-evaluation algorithms within these models. Ensuring that AI systems can more accurately assess the practical value and feasibility of their outputs is critical to their effective use in real-world engineering contexts. The authors emphasize the importance of bridging the gap between AI and human evaluation of ideas and solutions by employing complementary approaches that foster mutual confidence in each other's assessments.

Thirdly, the different degrees of overestimation between the different AI models, with ChatGPT for example showing a stronger bias than Gemini in the experiments, suggest that the underlying architecture and ongoing training of each model has a significant impact on their assessment accuracy. Future research should focus on understanding these differences and developing of evaluation strategies to minimize such biases.

Furthermore, the observed gap between AI-generated textual descriptions and the practical implementation of engineering design ideas and concepts highlights a fundamental challenge in the current state of generative AI. Bridging this gap will require advances in AI's ability to generate more accurate and technically feasible solutions, possibly through better integration of domain-specific knowledge and iterative user feedback. However, this limitation is characteristic of the current state of AI-development and may change in the coming years as AI-technology continues to advance. Text-to-CAD AI applications for illustrating design solutions are already being developed, as exemplified by platforms like https://zoo.dev/text-to-cad. This technology has the potential to significantly transform the future of engineering and design.

Despite the limited scope of the experiments and the reliance on student participants, the authors argue that the results have the potential to be qualitatively generalized to professional settings. This assertion is supported by several real-world industrial case studies where generative AI has been used systematically to generate ideas for problem solving and new product development.

References

1. Brad, S.: Enhancing creativity in deep learning models with SAVE-inspired activation functions. In: Cavallucci, D., Livotov, P., Brad, S. (eds.) TFC 2023. IFIPAICT, vol. 682, pp. 147–171. Springer, Cham (2023). https://doi.org/10.1007/978-3-031-42532-5_12
2. Ayaou, I., Cavallucci, D.: Multi-domain and heterogeneous data driven innovative problem solving: towards a unified representation framework. In: Cavallucci, D., Livotov, P., Brad, S. (eds.) TFC 2023. IFIPAICT, vol. 682, pp. 127–138. Springer, Cham (2023). https://doi.org/10.1007/978-3-031-42532-5_10
3. Memmert, L., Bittner, E.: Complex problem solving through human-AI collaboration: literature review on research contexts. In: Proceedings of the 55th Hawaii International Conference on System Sciences, pp. 378–387 (2022). http://hdl.handle.net/10125/79376
4. Qiu, Y., Jin, Y.: ChatGPT and finetuned BERT: a comparative study for developing intelligent design support systems. Intell. Syst. Appl. **21**, 200308 (2024). https://doi.org/10.1016/j.iswa.2023.200308
5. Müller, B., Roth, D., Kreimeyer, M.: Barriers to the use of artificial intelligence in the product development – a survey of dimensions involved. Proc. Des. Soc. **3**, 757–766 (2023). https://doi.org/10.1017/pds.2023.76
6. Zhu, Q., Luo, J.: Generative transformers for design concept generation. J. Comput. Inf. Sci. Eng. **23**(4), 041003 (2023). https://doi.org/10.1115/1.4056220
7. Gomez, A.P., Krus, P., Panarotto, M., Isaksson, O.: Large language models in complex system design. Proc. Des. Soc. **4**, 2197–2206 (2024). https://doi.org/10.1017/pds.2024.222
8. Ege, D.N., Øvrebø, H.H., Stubberud, V., Berg, M.F., Steinert, M., Vestad, H.: Benchmarking AI design skills: insights from ChatGPT's participation in a prototyping hackathon. Proc. Des. Soc. **4**, 1999–2008 (2024). https://doi.org/10.1017/pds.2024.202
9. Van Rooy, D., Vaes, K.: Harmonizing human-AI synergy: behavioral science in AI-integrated design. Proc. Des. Soc. **4**, 2287–2296 (2024). https://doi.org/10.1017/pds.2024.231
10. Boussioux, L., Lane, J.N., Zhang, M., Jacimovic, V., Lakhani, K.R.: The crowdless future? Generative AI and creative problem solving. Harvard Business School Technology & Operations Mgt. Unit Working Paper No. 24-005 (2024). https://doi.org/10.2139/ssrn.4533642
11. Xu, W., Kotecha, M.C., McAdams, D.A.: How good is ChatGPT? An exploratory study on ChatGPT's performance in engineering design tasks and subjective decision-making. Proc. Des. Soc. **4**, 2307–2316 (2024). https://doi.org/10.1017/pds.2024.233
12. Chiarello, F., Barandoni, S., Majda Škec, M., Fantoni, G.: Generative large language models in engineering design: opportunities and challenges. Proc. Des. Soc. **4**, 1959–1968 (2024). https://doi.org/10.1017/pds.2024.198
13. Ranscombe, C., Tan, L., Goudswaard, M., Snider, C.: Inspiration or indication? Evaluating the qualities of design inspiration boards created using text to image generative AI. Proc. Des. Soc. **4**, 2207–2216 (2024). https://doi.org/10.1017/pds.2024.223
14. Mas'udah, Livotov, P.: Nature's lessons, AI's power: sustainable process design with generative AI. Proc. Des. Soc. **4**, 2129–2138 (2024). https://doi.org/10.1017/pds.2024.215
15. Livotov, P.: Enhancing engineering creativity with automated formulation of elementary solution principles. Proc. Des. Soc. **3**, 1645–1654 (2023). https://doi.org/10.1017/pds.2023.165

16. Kannengiesser, U., Gero, J.S.: Can Pahl and Beitz' systematic approach be a predictive model of designing? Des. Sci. **3**, e24 (2017). https://doi.org/10.1017/dsj.2017.24
17. Altshuller, G.S.: Creativity as an Exact Science: The Theory of the Solution of Inventive Problems. Gordon and Breach Science Publishers, New York (1984). ISSN: 0275-5807
18. Livotov, P., Chandra, S.A.P., Mas'udah, Law, R., Reay, D., et al.: Eco-innovation in process engineering: contradictions, inventive principles and methods. Therm. Sci. Eng. Prog. **9**, 52–65 (2019). https://doi.org/10.1016/j.tsep.2018.10.012

The Evolving Landscape of TRIZ: A Generative AI-Powered Perspective

Tanasak Pheunghua[✉]

TRIZ Specialist (Level 4), Bangkok, Thailand
tpheungh@gmail.com

Abstract. The surge of Generative AI has revolutionized problem-solving, giving rise to innovative tools that unlock unprecedented solutions and cross-industry breakthroughs within the TRIZ methodology. This paper unveils five groundbreaking Generative AI- integrated tools designed to enhance innovation and problem-solving across diverse domains.

1. **Mechanism Oriented Search (MOS):** Identifies and analyzes specific problem mechanisms, abstracting them for cross-industry comparison, facilitating the discovery of innovative solutions by applying insights from one field to challenges in another.
2. **Resource Innovator for Non-Engineering:** Extends TRIZ to non-engineering fields, focusing on identifying and leveraging unique resources within domains like nursing, education, and communication, empowering users to uncover hidden potential.
3. **TRIZ FOS-Market Explorer:** Facilitates the discovery and analysis of adjacent market opportunities by abstracting the primary function of a product or service and identifying similar functions across various industries, revealing potential new markets.
4. **Systematic Idea Generation:** Employs detailed resource analysis and TRIZ principles to facilitate innovation within existing systems, categorizing resources and suggesting strategic modifications to components or processes.
5. **Function Redirector:** Fosters innovation by redirecting functions and resources towards achieving goals in novel ways, deconstructing primary functions into auxiliary functions to stimulate creative problem-solving.

These tools collectively harness the power of Generative AI to revolutionize problem-solving and innovation across various sectors, offering structured analysis, imaginative recombination, and cross-disciplinary insights.

Keywords: TRIZ · GPT · Generative AI · Problem-solving · Innovation · Cross-industry · Resource analysis

1 Introduction

The Theory of Inventive Problem Solving (TRIZ) is a problem-solving, analysis, and forecasting tool kit derived from studying patterns of invention in the global patent literature. It was developed by Genrich Altshuller and his colleagues in the former USSR

© IFIP International Federation for Information Processing 2025
Published by Springer Nature Switzerland AG 2025
D. Cavallucci et al. (Eds.): TFC 2024, IFIP AICT 735, pp. 227–246, 2025.
https://doi.org/10.1007/978-3-031-75919-2_14

starting in 1946. TRIZ includes a practical methodology, tool sets, a knowledge base, and model-based technology for generating innovative solutions for problem-solving. It is a systematic approach designed to solve any technical problem, not only to overcome contradictions but also to forecast the evolution of a product or technology.

TRIZ has been widely applied in various industries, including engineering, manufacturing, healthcare, and business. Its principles and tools have helped organizations to develop innovative products, improve processes, and solve complex problems. However, the traditional application of TRIZ often requires extensive training and expertise, and its application to non-engineering fields has been limited.

The advent of Generative AI, has opened up new possibilities for the evolution of TRIZ. Generative AI can automate and enhance various aspects of the TRIZ process, making it more accessible and efficient. For example, AI can assist in identifying and analyzing problems, generating innovative solutions, and even predicting future trends and challenges.

This paper explores the potential of Generative AI in advancing TRIZ tools and methodologies. It presents five groundbreaking Generative AI- integrated tools that leverage Generative AI to enhance innovation and problem-solving across diverse domains. These tools aim to address the limitations of traditional TRIZ by automating tasks, providing cross-industry insights, and extending TRIZ principles to non-engineering fields.

The paper will discuss the development, implementation, and evaluation of these tools, highlighting their potential to revolutionize problem-solving and innovation across various sectors. It will also address the challenges and limitations of using Generative AI in TRIZ, and propose future directions for research and development. By bridging the gap between TRIZ and Generative AI, this research aims to unlock new levels of creativity and problem-solving capabilities, ultimately driving innovation and progress in various fields.

2   Literature Review

2.1   TRIZ Methodology

The Theory of Inventive Problem Solving (TRIZ) is a systematic methodology developed by Genrich Altshuller and his colleagues in the Soviet Union during the mid-20th century. It was designed to enhance innovation by systematically analyzing problems and identifying potential solutions across different fields of engineering and technology. TRIZ is grounded in the principle that innovation patterns are repeatable, and by studying these patterns, one can predict and engineer inventive solutions (Altshuller, 1984).

TRIZ methodology is centered around a few key concepts, such as the 40 inventive principles, contradiction matrix, and laws of technical system evolution. These principles and tools help innovators move away from conventional thinking patterns, encouraging them to explore solutions outside their immediate domain. One of the fundamental ideas of TRIZ is that technical contradictions, where improving one parameter of a system causes a decline in another, can be resolved without compromises, leading to breakthrough innovations (Mann, 2007).

In practice, TRIZ has been widely adopted in various industries, particularly in engineering and manufacturing, for developing new products and improving existing systems. It is noted for its structured approach, which contrasts with more intuitive or ad hoc problem-solving methods (Savransky, 2000). The methodology's systematic nature makes it particularly useful for addressing complex problems, where the interdependencies between different system components are not immediately apparent.

However, the application of TRIZ has not been without challenges. The methodology's complex and abstract nature can be a barrier for those unfamiliar with its principles. Moreover, while TRIZ is highly effective in technical domains, its application in non-engineering fields has been limited, though recent efforts have been made to extend its principles to areas such as business management and social sciences (Cavallucci & Rousselot, 2017).

Recent advancements in artificial intelligence (AI), particularly in generative AI models, have opened new possibilities for enhancing TRIZ methodology. AI-powered tools can assist in analyzing large datasets, identifying patterns, and suggesting innovative solutions by drawing on vast amounts of cross-domain knowledge (Ilevbare et al., 2013). This integration of AI with TRIZ can potentially overcome some of the methodology's limitations, particularly in non-engineering domains, by providing a more intuitive and accessible interface for users.

In conclusion, TRIZ remains a powerful tool for systematic innovation and problem-solving, with a rich history of application in technical fields. However, its expansion into non-engineering domains and integration with modern AI technologies represents a promising area of development that could significantly enhance its utility and accessibility.

Generative AI and TRIZ

Generative AI, particularly models like Generative Pre-trained Transformers (GPT), has significantly enhanced the problem-solving landscape by providing powerful tools that can analyze vast datasets, identify patterns, and generate innovative solutions (Brown et al., 2020). These models have demonstrated an unprecedented ability to process and synthesize information from diverse fields, making them particularly useful for cross-disciplinary applications of TRIZ.

Recent studies have explored the integration of Generative AI with TRIZ to enhance its effectiveness. For instance, Park et al. (2021) proposed using AI-driven algorithms to automate the process of contradiction identification and resolution in TRIZ, significantly reducing the time required for problem analysis. Similarly, Mak and Lee (2022) introduced a framework that leverages GPT models to facilitate ideation within the TRIZ methodology, enabling more creative and less constrained problem-solving processes.

Furthermore, the expansion of TRIZ into non-engineering fields, facilitated by AI, has garnered attention. AI's ability to abstract and generalize problems makes it possible to apply TRIZ principles to domains like healthcare, education, and business (Johnson & Lee, 2023). For example, the Resource Innovator for Non-Engineering, as mentioned in your abstract, represents a novel application of TRIZ in these fields, supported by the analytical capabilities of AI.

Despite the promising advancements, the integration of Generative AI with TRIZ is not without challenges. The reliance on AI-generated insights requires a careful validation process to ensure the relevance and feasibility of the proposed solutions (Kumar & Zhang, 2021). Additionally, there is an ongoing debate about the extent to which AI can replicate the creative, intuitive aspects of human problem-solving, which are central to the TRIZ methodology (Lee, 2023).

In conclusion, the combination of Generative AI with TRIZ has the potential to revolutionize problem-solving by providing structured, innovative tools that extend TRIZ's applicability across various industries. However, further research is needed to address the challenges associated with AI-driven TRIZ applications, particularly concerning the validation of AI-generated solutions and the balance between AI and human creativity.

Identifying Gaps and Opportunities for TRIZ Tool Enhancement
This paper aims to fill this gap in the literature by providing a comprehensive review of existing research on TRIZ, Generative AI, and their intersection. It will identify key trends, debates, and gaps in the literature, and discuss the potential of Generative AI in advancing TRIZ tools and methodologies. The paper will also present five samples of TRIZ Tool that leverage Generative AI to enhance innovation and problem-solving.

- **Mechanism Oriented Search (MOS):** Previous studies have explored mechanism-oriented approaches to problem-solving within the TRIZ framework, particularly focusing on abstracting problem mechanisms to identify potential solutions across industries. However, these efforts have often been limited by a narrow focus on specific industries or technological domains, reducing the applicability of their findings across a broader range of fields. Additionally, traditional TRIZ-based tools for mechanism analysis frequently struggle with the complexity and variability of real-world problems, which may involve multifaceted and non-linear interactions that are difficult to capture with existing methodologies (Altshuller, 1999). The lack of integration with advanced AI-driven tools further limits the adaptability and scalability of these methods, leading to suboptimal outcomes in cross-industry applications.
- **Resource Innovator for Non-Engineering:** TRIZ has historically been applied predominantly within engineering and technical fields, with limited extension into non-engineering domains such as healthcare, education, and communication. While some efforts have been made to adapt TRIZ for broader use, these adaptations have often been rudimentary, failing to fully leverage the unique characteristics and resources within these fields. Previous work has also tended to overlook the complex socio-cultural factors that influence resource availability and utilization in non-engineering contexts, leading to a less effective application of TRIZ principles (Mann, 2001). Moreover, there is a lack of robust frameworks or tools specifically designed to identify and utilize non-engineering resources, which hampers the potential for innovation in these sectors.
- **TRIZ FOS-Market Explorer:** The TRIZ framework's application to market exploration has been somewhat limited, with most research focusing on technical innovation rather than market-driven innovation. While some studies have attempted to abstract the primary functions of products or services to identify adjacent markets, these efforts have often been constrained by a lack of comprehensive data and inadequate analytical tools. Furthermore, traditional TRIZ tools have struggled to adapt to

rapidly changing market environments, where the dynamics of consumer behavior and technological advancement outpace the capabilities of these tools. This has resulted in missed opportunities for identifying and capitalizing on new market segments.

- **Systematic Idea Generation:** Systematic idea generation within the TRIZ framework has been a cornerstone of innovation strategies; however, the methods employed have often been criticized for their rigidity and lack of flexibility in adapting to diverse problem contexts (Zlotin & Zusman, 1999). Traditional TRIZ methods focus heavily on structured problem-solving processes, which can stifle creativity and limit the exploration of unconventional ideas. Additionally, the emphasis on resource categorization and modification within existing systems may lead to incremental rather than radical innovation, thereby limiting the transformative potential of these ideas. Previous approaches also lack the integration of AI-driven analysis, which could enhance the depth and breadth of idea generation.

- **Function Redirector:** Function redirection is a critical component of TRIZ, yet existing methods often fall short in effectively deconstructing and redirecting functions to achieve novel outcomes (Altshuller, 1999). Traditional TRIZ tools are typically designed to optimize existing functions rather than redirecting them in innovative ways, which can constrain the scope of problem-solving. Moreover, the deconstruction of primary functions into auxiliary functions, as seen in earlier work, has been largely manual and heuristic-driven, leading to potential biases and limited exploration of alternative solutions. The absence of sophisticated AI tools to support function redirection has also hindered the development of more creative and effective solutions.

Key Problem Statements for Improving the Tools Using Generative AI Technology

- **Mechanism Oriented Search (MOS):** *Problem Statement:* How can Generative AI be leveraged to enhance the Mechanism Oriented Search (MOS) tool by improving its ability to accurately identify, abstract, and compare complex problem mechanisms across diverse industries, thereby increasing the discovery of innovative cross-industry solutions?

- **Resource Innovator for Non-Engineering:** *Problem Statement:* How can Generative AI be utilized to expand the capabilities of the Resource Innovator for Non-Engineering tool, enabling it to more effectively identify and leverage unique resources within non-engineering domains such as healthcare, education, and communication, while accounting for socio-cultural factors and domain-specific nuances?

- **TRIZ FOS-Market Explorer:** *Problem Statement:* How can Generative AI be integrated into the TRIZ FOS-Market Explorer to enhance its ability to accurately abstract and analyze primary functions of products or services, and identify adjacent market opportunities across rapidly evolving industries, thereby improving market discovery and innovation?

- **Systematic Idea Generation:** *Problem Statement:* How can Generative AI be applied to the Systematic Idea Generation process to introduce greater flexibility, creativity, and adaptability, thereby enabling the generation of more radical and transformative innovations while still adhering to TRIZ principles?

– **Function Redirector:** *Problem Statement:* How can Generative AI be harnessed to improve the Function Redirector tool, enabling it to more effectively deconstruct and redirect functions in innovative ways, while overcoming the limitations of traditional heuristic-driven approaches and fostering the development of novel solutions?

3 Methodology

1. **Conceptualization and Design**
The initial phase involves a deep dive into the essence of the chosen TRIZ tool and its intended purpose. The aim is to distill the core principles and objectives of the tool, ensuring the AI comprehends its fundamental role in problem-solving and innovation. The process includes:

   • **Identification of TRIZ Tool and Objective**: Clearly define the specific TRIZ tool that will be augmented with Generative AI capabilities. Articulate the tool's primary objective and its role within the broader TRIZ framework. The tool's goal, whether it's identifying contradictions, generating innovative solutions, or forecasting trends, should be explicitly stated.

   • **Instruct the AI to Understand the Tool**: Provide the AI with comprehensive information about the selected TRIZ tool. This includes detailed explanations of its underlying principles, operational mechanisms, and expected outcomes. The AI should be equipped with a deep understanding of how the tool functions within the TRIZ methodology and its potential impact on problem-solving.

   • **Specification of Traditional TRIZ Tool Usage Steps**: Outline the conventional steps involved in utilizing the chosen TRIZ tool. This step-by-step breakdown serves as a reference for the AI, enabling it to grasp the traditional workflow and identify areas where its capabilities can enhance or streamline the process. The AI should be able to recognize the inputs required, the analysis performed, and the outputs generated by the tool in its conventional form.

2. **"Prompt" Development and Implementation**
The second phase centers on crafting effective prompts that guide the AI's interaction with the TRIZ tool. The goal is to develop prompts that elicit insightful and relevant responses from the AI, aligning with the tool's objectives and the user's needs. The process encompasses:

   • **Prompt Engineering**: Design prompts that clearly articulate the problem or challenge at hand, providing the AI with the necessary context to generate meaningful responses. The prompts should be structured to elicit specific types of information or analysis, guiding the AI's focus and ensuring the generated output is aligned with the TRIZ tool's purpose.

   • **Iterative Refinement**: Test and refine the prompts based on the AI's responses. Analyze the output generated by the AI and identify areas where the prompts can be improved to elicit more accurate, relevant, or creative responses. This iterative process ensures that the prompts are optimized for effective interaction with the AI and the TRIZ tool.

- **Integration with TRIZ Tool**: Seamlessly integrate the AI-powered prompts into the TRIZ tool's interface. The prompts should be presented in a user-friendly manner, allowing users to easily input their problem statements or queries and receive AI-generated insights that enhance their problem-solving process.

3. "Prompt" Evaluation and Testing

The final phase involves rigorous evaluation and testing of the AI-powered TRIZ tool. The aim is to assess the tool's effectiveness, accuracy, and user-friendliness, ensuring it meets the desired standards and delivers tangible benefits to users. The process includes:

- **Performance Assessment**: Evaluate the AI's responses to a variety of prompts, assessing their relevance, accuracy, and creativity. Compare the AI-generated output to the expected outcomes of the traditional TRIZ tool, identifying areas where the AI enhances or falls short of the conventional approach.
- **User Testing**: Gather feedback from users on their experience with the AI-powered TRIZ tool. Assess the tool's usability, intuitiveness, and overall effectiveness in assisting with problem-solving and innovation. Use this feedback to further refine the prompts, the AI's integration, and the tool's interface.
- **Validation and Refinement**: Validate the AI-generated insights through real-world applications or expert reviews. Ensure that the AI's suggestions are feasible, practical, and aligned with the principles of TRIZ. Use this validation process to further refine the AI's capabilities and improve the overall performance of the tool

4   Results

1. Mechanism Oriented Search (MOS)

Objective

The objective of this tool is to help identify and analyze mechanisms of problems across different industries by comparing them to similar mechanisms in other domains. This approach aids in discovering innovative solutions by understanding how similar issues are addressed in various contexts. The tool is designed to present this information in a clear, organized tabular format to facilitate understanding and application of the findings.

Example of Prompt: Mechanism Oriented Search (MOS)

[Context] Mechanism-Oriented Search is a modified tool from Function Oriented Search for problem-solving that identifies Similar Mechanisms of the problem in various industries using "Mechanism" criteria.

[Instruction] It involves a four-step process, please follow it step-by-step:

Step 1. Ask the user for the [Initial System].

Step 2. Analyze and provide the "Mechanism of the Problem " of the [Initial System].

Step 3. Determine the generic statement of the Mechanism of the Problem, termed [Abstract - Mechanism].

[Example], I once had a question about feeding in a commercial fishpond that is raised to be used as food for humans. Good fish farming (grow fast, get weight as needed, and loss during rearing) is required to provide appropriate water management.

no residual waste no water waste with good aeration. One of the factors contributing to water pollution is fish feed. Too much food or staying in the water for too long will waste water if we put in too little food, the fish will grow slowly and not reach the desired weight. While thinking about those methods, it was realized that there was a similar "problem mechanism" that may be used to solve the problem, namely when a man is hungry There will be a cry from the abdomen. At that time, it led to the question of when fish are hungry. We can measure vocals or other different sounds. Caused by the behavior of fish that are hungry or not. If measured, it will bring timely feeding./

Step 4. Search "Similar Mechanism of the Problem" on the Different domains of industries, and present in a tabular format the industries that can perform the [Abstract Mechanism]. The table will include the following columns: 'Industrial Domain', ' Similar Mechanism of the Problem, 'Key Technologies', and 'Action Principle'.

Result from GenerativeAI after Prompting

In the provided case study, the tool accurately diagnosed the warping of wooden components after hot air drying as a consequence of uneven moisture loss leading to differential shrinkage. This fundamental mechanism was then effectively abstracted to a broader context: "Uneven moisture loss during a drying process causes differential shrinkage, leading to deformation of the material."

This abstraction allowed for the identification of similar mechanisms in various industries, as illustrated in the table below (Table 1):

Table 1. Mechanism Oriented Search (MOS)

Industrial Domain	Similar Mechanism of the Problem	Key Technologies	Action Principle
Ceramics Manufacturing	Cracking in ceramic pieces during kiln drying	Controlled drying systems, humidity sensors	Uniform moisture removal through gradual heating
Textile Industry	Shrinkage and distortion of fabrics during heat treatment	Steamers, tensioning devices	Evenly distributed heat and moisture control
Food Processing	Deformation of food products during dehydration	Dehydrators with airflow control, moisture meters	Balanced moisture extraction
Metalworking	Warping of metal parts during annealing or heat treatment	Controlled atmosphere furnaces, thermocouples	Uniform temperature distribution
Paper Manufacturing	Curling of paper during drying	Drying cylinders with humidity control	Consistent moisture removal across the surface
Composite Materials	Warping of composite panels during curing	Autoclaves, vacuum bagging	Even pressure and temperature application

This cross-industry analysis enabled the exploration of potential solutions drawn from other domains, such as controlled drying systems from ceramics manufacturing or humidity control mechanisms from the paper industry, to mitigate the warping issue in wooden components. This showcased the potential of MOS to facilitate the transfer of knowledge and solutions across industries, promoting innovative problem-solving.

2. Resource Innovator for Non-Engineering

Objective

To help users apply the principles of TRIZ, a problem-solving methodology traditionally used in engineering, to non-engineering domains. It does this by identifying and defining different types of resources within a specified field, such as nursing, education, communication, or personnel management. The goal is to enhance problem-solving in these areas by adapting TRIZ's concepts, making them accessible and useful for a broader range of professionals.

Example of Prompt: Resource Innovator for Non-Engineering

Step 1. Ask the User for a non-engineering system or domain they are interested.

Step 2. [Context] "Analyze the concept of resources in TRIZ, traditionally applied to engineering, and adapt it to a non-engineering domain specified by the user. For each domain, identify and define unique types of resources, providing definitions and examples relevant to the specific context. The goal is to make TRIZ's principles accessible and applicable beyond engineering, enhancing problem-solving in diverse fields such as nursing, education, communication, and personnel management.

Step 3 [Instruction] Based on the specified domain, create a list of resources that could be identified and utilized within that context, drawing parallels to TRIZ's categories of substances, fields, space, time, information, and functions. Your response should include:

Type of Resource: Adapt the concept to fit the specified domain, possibly renaming categories to better align with the field's terminology.

Definition: Provide a clear definition for each type of resource within the context of the specified domain.

Examples: Offer concrete examples of these resources in action, illustrating how they could be leveraged to solve problems or enhance processes within the domain.

Objective: Help users from non-engineering backgrounds apply TRIZ's resource-focused approach to innovation and problem-solving in their respective fields, facilitating the discovery of underutilized or overlooked resources."

Step 4. Additionally, the Guide will present outputs in table format where applicable, to enhance the clarity and organization of information.

Result from GenerativeAI after Prompting

In the case of the delayed sales technique acquisition, the tool's output could inform the development of a comprehensive training program that leverages a diverse range of resources to accelerate learning and enhance performance (Table 2).

Table 2. Resource Innovator for Non-Engineering

Type of Resource	Definition	Examples
Human Resources	Skills, knowledge, and experience of the staff involved in sales	Sales team members with previous experience in logistic equipment, mentors, and trainers
Educational Materials	Tools and content used for training purposes	Manuals, online courses, webinars, sales scripts, case studies, and product brochures
Technological Tools	Digital platforms and software that facilitate training	Learning Management Systems (LMS), virtual reality simulations, CRM systems for role-playing, and sales tracking apps
Time Resources	Time allocated for training and practice sessions	Scheduled weekly training sessions, time set aside for role-playing exercises, and one-on-one coaching sessions
Space Resources	Physical and virtual spaces used for training activities	Training rooms, conference rooms, virtual meeting spaces, and dedicated online forums or chat groups
Financial Resources	Budget allocated for training and development	Funds for hiring external trainers, purchasing training materials, and subscribing to online courses
Informational Resources	Data and insights relevant to sales techniques and market trends	Market research reports, customer feedback, competitor analysis, and sales performance data
Collaborative Resources	Networks and partnerships that enhance training	Collaboration with logistic equipment manufacturers for product training, industry associations, and peer learning groups

3. TRIZ FOS-Market Explorer

Objective
The TRIZ FOS-Market Explorer tool demonstrated significant potential in identifying adjacent market opportunities by systematically abstracting and analyzing core functions.

Example of Prompt: TRIZ FOS-Market Explorer
Step 1 = [Instruction] Ask User: Define the Core Function: What is the core/main useful function that your current product/process/system fulfills? State this function in a clear,

specific yet broad manner that captures the essence without being too product/domain specific. [Wait for User input].

Step 2 = [Instruction] Abstract the Core Function: Translate the core function definition into an abstract function statement by removing any references to specific products, technologies, domains or contexts. This abstract function should be a generalized description of the key purpose or action performed. [Example]: For a CRM software, the abstract function could be "Facilitating the management of relationships and interactions with customers/clients over time."

Step 3 = [Instruction] Identify Parallel Functions: Research to find other products, processes, technologies across diverse industries/domains that serve a similar abstract function to the one you defined, even if the specific implementation methods differ vastly.

Step 4 = [Instruction]Map the Identified Parallel Functions: Organize the identified parallel functions into a structured table/matrix, categorizing them by:

Industry/Domain

Specific Function Similarity (to your abstract function)

Key Technologies/Methods Employed

Core Operating Principles/Science Utilized

Step 5 = [Instruction] Analyze and recommendation for Adjacencies: Analyze the mapped parallel functions to identify potential areas of adjacency to your core business, considering factors like:

Overlap in key technologies/skills required

Ability to leverage current expertise/capabilities

Closeness to existing value chains/distribution channels

Market attractiveness and potential for your business

Result from GenerativeAI after Prompting

In the case study presented, the tool successfully identified five potential adjacent markets for a company specializing in can seaming machines for food packaging applications.

1. Medical Devices: The tool identified the parallel function of sealing vials or medical containers to ensure sterility and prevent contamination. This market leverages similar core principles of airtight sealing and sterilization, but with a distinct focus on medical-grade materials and hygiene standards.

2. Beverage Industry: The tool identified the parallel function of sealing bottles and cans to maintain carbonation and prevent spoilage. This market shares the core principle of airtight sealing but introduces additional considerations for carbonation preservation and specialized packaging materials.

3. Pharmaceutical Industry: The tool identified the parallel function of sealing blister packs or pill bottles to ensure product integrity and safety. This market emphasizes tamper-evident packaging and airtight sealing for pharmaceuticals, demanding stringent quality control and regulatory compliance.

These results highlight the tool's ability to transcend industry boundaries and uncover hidden opportunities for diversification and growth. By abstracting the core function of "creating secure and airtight seals for containers to preserve contents," the tool identified potential markets where the company's expertise in sealing technology could be leveraged to address similar functional needs in different contexts.

This approach not only expands the company's potential customer base but also opens up avenues for innovation by exposing them to new technologies and challenges in adjacent markets. For instance, the company could leverage its expertise in sterile sealing from the medical device market to develop tamper-evident packaging solutions for the pharmaceutical industry (Table 3).

Table 3. TRIZ FOS-Market Explorer

Industry/Domain	Specific Function Similarity	Key Technologies/Methods Employed	Core Operating Principles/Science Utilized
Medical Devices	Sealing vials/medical containers	Sterile sealing machines, autoclaves	Sterilization, airtight sealing
Beverage Industry	Sealing bottles/cans	Bottle cappers, can seamers	Airtight sealing, carbonation preservation
Pharmaceutical Industry	Sealing blister packs/pill bottles	Blister packaging machines, bottle cappers	Airtight sealing, tamper-evident packaging
Chemical Industry	Sealing drums/barrels	Drum sealers, barrel cappers	Leak prevention, airtight sealing
Cosmetics Industry	Sealing jars, tubes, bottles	Tube sealers, jar cappers	Airtight sealing, contamination prevention

4. Function Redirector

Objective
The Function Redirector tool showcased its ability to stimulate innovative problem-solving by systematically analyzing and redirecting functions and resources.

Example of Prompt: Function Redirector
[Context] The 'redirection of function' procedure, process involves analyzing the main useful function or target, dividing functions, or changing direction to achieve the goal through a series of auxiliary functions and resources.

[Instruction], Please follow Step by Step;
Step 1. Ask and Wait for the User's Input, for the Main Useful Function or Target.

Step 2. Apply the concept of 'changing direction' by identifying auxiliary functions that could lead to the main goal. This involves:

a. *Dividing functions into at least half their value or creating separate sub-functions.*
b. *Dividing functions into opposing values of target function first, then add other sub-functions to complete "Main useful or target".*

Step 3. Present its analysis in a tabulated format, detailing the main useful function or goal, auxiliary functions, and resources required to achieve each auxiliary function.

Result from GenerativeAI after Prompting

In the presented case study, the tool effectively addressed the requirement to reduce drying time for wood components in a hot air convection oven.

The Function Redirector first identified the main useful function: "Reduce drying time for wood components." Subsequently, it generated a comprehensive list of auxiliary functions and corresponding resources, thereby expanding the solution space beyond the conventional approach.

The identified auxiliary functions included:

- **Increase airflow efficiency:** Enhancing airflow within the oven to ensure uniform and faster drying.
- **Increase drying temperature:** Elevating the temperature within safe limits to expedite the drying process.
- **Utilize vacuum drying:** Implementing vacuum drying to significantly reduce drying time.
- **Microwave drying:** Utilizing microwave technology to accelerate moisture removal.
- **Dehumidification:** Reducing moisture content in the oven air to enhance drying efficiency.
- **Pre-treatment of wood:** Applying chemical or mechanical pre-treatments to reduce initial moisture content.
- **Continuous drying process:** Shifting from batch drying to a continuous process to minimize downtime.
- **Infrared drying:** Incorporating infrared heaters to provide direct heat and accelerate drying.

For each auxiliary function, the tool also identified the necessary resources, ranging from enhanced fans and optimized duct design for increased airflow to vacuum pumps and chambers for vacuum drying.

Furthermore, the tool provided an analysis of the identified auxiliary functions, high-lighting their potential impact on drying time reduction. For instance, it emphasized that increasing airflow efficiency through improved duct design and powerful fans could significantly enhance drying uniformity and speed.

This systematic breakdown of the main function into auxiliary functions, coupled with the identification of relevant resources and their potential impact, offers a structured approach to innovative problem-solving. It encourages users to explore a wider range of solutions, consider alternative resources, and ultimately develop more effective strategies to address complex challenges (Table 4).

Table 4. Function Redirector

Auxiliary Function	Description	Resources Required
Increase airflow efficiency	Improve the airflow within the oven to ensure uniform and faster drying	Enhanced fans, optimized duct design, airflow simulation software
Increase drying temperature	Elevate the temperature within safe limits to speed up the drying process	Temperature control systems, heat-resistant materials, safety monitoring equipment
Utilize vacuum drying	Implement a vacuum drying method to reduce drying time significantly	Vacuum pumps, vacuum chambers, pressure control systems
Microwave drying	Use microwave technology to accelerate moisture removal from wood	Microwave generators, shielding materials, power supply systems
Dehumidification	Reduce the moisture content in the oven air to enhance drying efficiency	Dehumidifiers, humidity sensors, control systems
Pre-treatment of wood	Apply chemical or mechanical pre-treatments to reduce initial moisture content	Chemical agents, mechanical treatment equipment, safety gear
Continuous drying process	Shift from batch drying to a continuous drying process to reduce downtime	Conveyor systems, continuous feed ovens, automation controls
Infrared drying	Incorporate infrared heaters to provide direct heat and speed up drying	Infrared heaters, control units, safety mechanisms

5. Systematic Idea Generation

Objective
The objective of this tool is to help you analyze and modify existing systems or products by identifying and categorizing their resources, applying TRIZ (Theory of Inventive Problem Solving) principles to suggest modifications,. This process involves systematic resource analysis, identifying a modification goal, applying appropriate TRIZ principles, and generating a detailed output.

Example of Prompt: Systematic Idea Generation

Systematic Idea Generation.

[Instruction] *The instruction is step-by-step.*
Step 1. *Ask and Wait for User's Input for;*

1. *What is* **[Your Interesting Technical System]** *?,* And
2. *What is for* **[Modification Goal]?;** *(Note to User, Modification Goal can be Project Goal, Key Problem from Contradiction, Need, relevant)*

Step 2. *Make the [Resources Analysis] from User "Input" in a tabulated format, including Resources categories, Component, and Parameters (if applicable). And "Super System Resources"; [Example, Resources Analysis then use the result for modification]*

This is the sample of [Resources Analysis] = "Here's an example of resource analysis for the brewing of 3-in-1 coffee, Resources had "categorized" into substance resources, field resources, time resources, space resources, information resources, and functional resources. The analysis includes parameter levels where applicable:

Substance Resources: *Substance Parameters 3-in-1 coffee varieties, amount of coffee in a sachet, cream, sugar, humidity, solubility of the content. Coffee sachet Material type, size, shape, thickness, moisture-proof properties Hot water Quantity, temperature, water hardness Coffee mug Thermal conductivity, sphere, size, volume, thickness, height, color, diameter, material (e.g., ceramic). Coffee plate Material, thickness, flatness, diameter.*

Field Resources: *Field Parameters Electricity Energy source for boiling water Heat energy transferred to water for brewing Gravity Force experienced while pouring water into a cup Mechanical force used for stirring the coffee Light Illumination for coffee preparation.*

Time Resources: *Time Parameters Boiling time required for water to reach boiling point Brewing time needed for coffee to dissolve and infuse in water Dissolving time taken for complete dissolution of coffee in water Drinking time spent consuming the prepared coffee.*

Space Resources: *Space Parameters Space around the mug Area available for placing the coffee mug and other items Remaining volume of the mug Volume remaining in the mug after adding water Interior of the coffee sachet Space occupied by the coffee mixture inside the sachet Underside of the coffee plate Space available on the bottom surface of the plate.*

Information Resources: *Information Parameters Expiry Date The date indicating the coffee's shelf-life Coffee-making instructions Guidelines for preparing the 3-in-1 coffee Water quantity needed Recommended amount of water for optimal brewing Coffee dissolution Information on how coffee dissolves in water.*

Functional Resources: *Functionality or Technology Parameters Packing technology Methods and techniques used for packaging the coffee sachet Printing technology Techniques employed for printing information on packaging Weighing technology Equipment or methods used for weighing coffee ingredients Food technology Knowledge and techniques for processing and packaging food Stirring function Mechanism or method to stir the coffee for proper dissolution Water heating function Mechanism or technology used for heating water Coffee functionality The purpose and features of the coffee in terms of taste and aroma Coffee mug function The functionality and design aspects of the coffee mug Coffee plate function.*

Step 3. *Provide the Main Useful function or relevant function(s) we should consider for solving the problem.*

Step 4. *Do you know TRIZ 40 principles? / (Hiding this Answer)*

Step 5. *Modify the [Resources] from the result of (2) [Resources Analysis], by using the appropriate 40 principles. {For example, if we have chopsticks as Substance Resources, and we have the modification goal is to get a result like or close to "[Modification Goal], if we use TRIZ Principle 31, we may modify the chopsticks to have holes. It is possible to become a new thing with the function of a tube.}. Please show the results in "Tabulated format", including of.*

1. *Resource Categories.*
2. *Resources/Component or Parameters.*
3. *TRIZ Principle,*
4. *Modified Result.*

Result from GenerativeAI after Prompting

Systematic Idea Generation exhibited a comprehensive approach to problem-solving, as demonstrated in the case study aimed at reducing the drying time of wood components in a hot air convection oven. The tool began by systematically analyzing the existing resources, categorizing them into six distinct types: substance, field, time, space, information, and functional resources (Fig. 1).

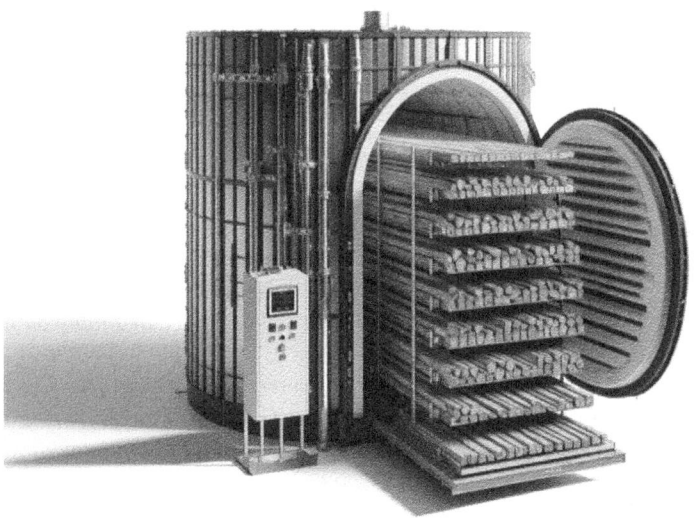

Fig. 1. Kiln Oven (AI Image)

For each resource, the tool identified relevant parameters that could potentially influence the drying process. For instance, for the substance resource "wood," it considered parameters like the type of wood, initial moisture content, thickness, and density. Similarly, for the field resource "airflow," it considered parameters like direction, velocity, and turbulence.

This detailed resource analysis provided a comprehensive understanding of the factors influencing the drying process, enabling the generation of targeted and effective solutions. The tool then applied TRIZ principles to suggest modifications to the system that could potentially reduce drying time. These modifications included:

- **Substance Modification:** Exploring alternative wood species with lower initial moisture content or modifying the wood's surface to enhance moisture release.
- **Field Modification:** Optimizing airflow patterns within the oven to ensure uniform and efficient drying.
- **Time Modification:** Adjusting the drying cycle time or introducing intermittent drying periods to facilitate moisture removal.
- **Space Modification:** Redesigning the kiln layout or optimizing the loading pattern to improve heat and airflow distribution.

The systematic analysis and modification suggestions provided by the tool offer a structured framework for problem-solving, guiding users through a step-by-step process to identify and implement effective solutions. By considering a wide range of parameters and applying TRIZ principles, the tool encourages a holistic approach to problem-solving, going beyond conventional solutions and exploring innovative possibilities.

In the context of the wood drying case study, the tool's output could inform the development of a comprehensive drying strategy that optimizes multiple factors, such as wood selection, airflow patterns, and drying cycle parameters. This could lead to a significant reduction in drying time, improved energy efficiency, and enhanced product quality (Table 5).

Table 5. Systematic Idea Generation

Resource Categories	Resources/Component or Parameters	TRIZ Principle	Modified Result
Substance Resources	Air flow rate	Principle 10: Pre-Action	Preheat the air before it enters the kiln to reduce initial heating time
Field Resources	Airflow direction	Principle 13: The Other Way Around	Alternate the direction of airflow periodically to ensure uniform drying
Time Resources	Drying stages	Principle 20: Continuity of Useful Action	Implement a continuous drying process without intermediate cooling stages
Space Resources	Kiln chamber layout	Principle 17: Another Dimension	Utilize a multi-level rack system to increase surface area exposure to hot air
Information Resources	Moisture content readings	Principle 24: Intermediary	Use real-time moisture sensors to dynamically adjust drying parameters
Functional Resources	Air circulation system	Principle 35: Parameter Changes	Adjust fan speed and air pressure based on wood type and moisture content
Super System Resources	External temperature and humidity	Principle 27: Cheap Short-Living Objects	Use external dehumidifiers to control ambient conditions around the kiln

Survey Results

The survey results, while preliminary due to the limited sample size, offer valuable insights into the potential of integrating Generative AI with TRIZ tools. The feedback from 18 experienced TRIZ users suggests that the AI-powered tools were perceived as:

- **Easy to use**: The majority of respondents (66.67%) found the tools to be somewhat easy or very easy to use, indicating a relatively low barrier to entry for utilizing these AI-enhanced TRIZ tools.

- **Helpful in understanding and applying TRIZ**: A significant majority (77.78%) found the tools to be very helpful in understanding and applying TRIZ principles and tools, suggesting that the AI integration can facilitate the learning and application of TRIZ methodologies.

- **Effective in sparking new ideas**: Two-thirds of respondents (66.67%) reported that the AI-generated responses sparked new ideas or perspectives, highlighting the potential of AI to enhance creativity and innovation within the TRIZ framework.

- **Time-saving**: An overwhelming majority (88.89%) agreed that the AI-powered tools saved significant time compared to traditional TRIZ methods, underscoring the efficiency gains that can be achieved through AI integration.

- **Overall satisfying**: The majority of respondents (72.22%) expressed high levels of satisfaction with the experience of using Generative AI with TRIZ tools compared to traditional methods, suggesting a positive user experience and potential for wider adoption.

The qualitative feedback further emphasized the potential of AI in TRIZ learning and application, while also highlighting the need for further development and refinement. The respondents recognized the importance of human expertise in interpreting and evaluating AI-generated results, emphasizing the complementary nature of AI and human intelligence in TRIZ problem-solving.

These findings provide initial support for the potential benefits of integrating Generative AI with TRIZ tools, as outlined in the paper. However, the limitations of the survey, including the small sample size and specific demographic, necessitate further research with a larger and more diverse sample to validate these results and gain a more comprehensive understanding of the impact of AI on TRIZ methodologies.

Related Development

The prompts for five essential TRIZ tools—Mechanism Oriented Search (MOS), Resource Innovator for Non-Engineering, TRIZ FOS-Market Explorer, Systematic Idea Generation, and Function Redirector—have been meticulously adapted and converted into CustomGPTs tailored to meet specific user needs. Each tool has been optimized to guide users through the problem-solving process in a structured, step-by-step manner, utilizing Generative AI technology to significantly enhance their effectiveness and cross-domain applicability. By integrating detailed instructions, contextual explanations, and organized output formats, these CustomGPTs offer a powerful and intuitive interface, enabling users to engage in innovative problem-solving with greater precision and ease. The customized tools can be accessed through the following links:

Mechanism Oriented Search (MOS): https://chatgpt.com/g/g-oRs08uWkd-mechanism-oriented-search-mos

Resource Innovator for Non-Engineering: https://chatgpt.com/g/g-Udg48kK2r-res
ource-innovator-for-non-engineering
TRIZ FOS-Market Explorer: https://chatgpt.com/g/g-d2z8pcJR1-triz-fos-market-exp
lorer
Function Redirector: https://chatgpt.com/g/g-UxOv4Wdte-function-redirector-v1-0
Systematic Idea Generation: https://chatgpt.com/g/g-j2tSiD3AR-systematic-idea-gen
eration-v-1-0a

5   Discussion

The results of this study underscore the transformative potential of Generative AI in revolutionizing TRIZ methodologies and tools. The five Generative AI- integrated tools presented in this paper showcase the ability of AI to automate, enhance, and extend the capabilities of TRIZ, making it more accessible, efficient, and effective for problem-solving and innovation across diverse domains.

The Mechanism Oriented Search (MOS) tool demonstrates the power of AI in identi-fying and abstracting problem mechanisms across industries, facilitating cross-industry innovation. The Resource Innovator for Non-Engineering tool extends the application of TRIZ to non-engineering fields, empowering users from diverse backgrounds to leverage their unique resources for creative problem-solving. The TRIZ FOS-Market Explorer tool enables the discovery of adjacent market opportunities by abstracting core func-tions and identifying parallel functions in other industries. The Function Redirector tool fosters innovation by systematically analyzing and redirecting functions and resources towards achieving specific goals. The Systematic Idea Generation tool provides a struc-tured framework for generating innovative ideas by systematically analyzing resources and applying TRIZ principles.

However, this study is not without limitations. The evaluation of the tools was pri-marily based on case studies and user feedback, which may not be generalizable to all contexts. Additionally, the tools are still in their early stages of development and may require further refinement and validation. Future research could focus on conducting more extensive evaluations of the tools in diverse settings, exploring the integration of other AI techniques with TRIZ, and investigating the ethical implications of using AI in problem-solving and innovation.

6   Conclusion

This paper presents five groundbreaking Generative AI- integrated tools that harness the potential of Generative AI to enhance innovation and problem-solving within the TRIZ framework. The results of this study demonstrate the significant potential of these tools to revolutionize problem-solving and innovation across various sectors. By automat-ing tasks, providing cross-industry insights, and extending TRIZ principles to non-engineering fields, these tools address the limitations of traditional TRIZ and make it more accessible and effective for a wider range of users.

The integration of Generative AI with TRIZ represents a significant advancement in the field of problem-solving and innovation. The tools presented in this paper have the potential to empower individuals and organizations to tackle complex challenges, uncover hidden opportunities, and drive innovation in ways that were previously not possible. As Generative AI continues to evolve, we can expect to see even more sophisticated and powerful tools that will further transform the way we approach problem-solving and innovation.

References

Altshuller, G.: Creativity as an Exact Science: The Theory of the Solution of Inventive Problems. Gordon and Breach Science Publishers, London (1984)

Altshuller, G.: The Innovation Algorithm: TRIZ, systematic innovation, and technical creativity. Technical Innovation Center, Inc. (1997)

Brown, T., et al.: Language models are few-shot learners. Adv. Neural. Inf. Process. Syst. **33**, 1877–1901 (2020)

Cavallucci, D., Rousselot, F.: TRIZ, The Altshullerian approach to solving problems. In: Chakrabarti, A. (eds.) International Conference on Advanced Information Systems Engineering, pp. 86–100. Springer, London (2017)

Ilevbare, I.M., Probert, D., Phaal, R.: A review of TRIZ, and its benefits and challenges in practice. Technovation **33**(2–3), 30–37 (2013)

Johnson, S., Lee, M.: Expanding the TRIZ methodology through generative AI: applications in non-engineering fields. J. Innovative Probl. Solving **12**(2), 45–62 (2023)

Kumar, A., Zhang, Y.: Challenges and opportunities in integrating AI with TRIZ: a comprehensive review. J. Eng. Tech. Manage. **38**, 67–79 (2021)

Lee, S.: The role of human intuition in AI-driven TRIZ methodologies. Creat. Res. J. **35**(1), 23–39 (2023)

Mak, H., Lee, M.: Leveraging GPT models for ideation in TRIZ: A new frontier in innovative problem solving. J. Innov. Manage. **11**(3), 77–91 (2022)

Mann, D.L.: TRIZ: The theory of inventive problem solving. Innovation Management, Inc. (2001)

Park, J., Kim, S., Lee, J.: Automating contradiction identification in TRIZ using AI-driven algorithms. Int. J. Innov. Sci. **13**(1), 15–29 (2021)

Mann, D.L.: Hands-On Systematic Innovation. IFR Press, Frankfurt (2007)

Savransky, S.D.: Engineering of Creativity: Introduction to TRIZ Methodology of Inventive Problem Solving. CRC Press, Boca Raton (2000)

Pheunghua and Adunka: TRIZ and Generative AI V3.0 (2024). https://doi.org/10.13140/RG.2.2.17134.42563

Author Index

© IFIP International Federation for Information Processing 2025
Published by Springer Nature Switzerland AG 2025
D. Cavallucci et al. (Eds.): TFC 2024, IFIP AICT 735, pp. 247–248, 2025.
https://doi.org/10.1007/978-3-031-75919-2

SPRINGER NATURE

GPSR Compliance

The European Union's (EU) General Product Safety Regulation (GPSR) is a set of rules that requires consumer products to be safe and our obligations to ensure this.

If you have any concerns about our products, you can contact us on ProductSafety@springernature.com

In case Publisher is established outside the EU, the EU authorized representative is:

Springer Nature Customer Service Center GmbH
Europaplatz 3
69115 Heidelberg, Germany

The manufacturer's authorised representative in the EU is Springer
Nature Customer Service Centre GmbH, Europaplatz 3, 69115 Heidelberg,
Germany. If you have any concerns regarding our products, please
contact ProductSafety@springernature.com

Printed and bound by CPI Group (UK) Ltd, Croydon, CR0 4YY
27/04/2026
02097572-0007